Suckled Calf Production

WITHDRAWN

Second Edition

Richard Fuller ARAgS

Manager, JSR Farming Group Ltd, Givendale
Pocklington, York, England

First published in Great Britain by
Chalcombe Publications
Painshall, Church Lane, Welton
Lincoln LN2 3LT

First edition 1998
Second edition 1998

© R Fuller 1988, 199

ISBN 0 948617 40 3

Printed in Great Britain by Ruddocks Colour Printers, Lincoln.

Contents

Preface

In 1974 I arrived at Givendale Farm, on the edge of the Yorkshire Wolds, to develop a suckled calf enterprise based on 131 hectares of unploughable grassland on the farm.

I recognised that most lowland farmers viewed suckled calf production as a 'Cinderella' enterprise. Poor returns were often the consequence of insufficient management control of the suckler cow. However, I was confident that, with effective supervision, a suckler herd would make a useful contribution to farm income.

The management techniques which I have developed at Givendale have been rewarded by significant, if not dramatic improvements in herd performance. As a result a great deal of interest has been shown in our suckler herds. I have hosted many farm visits showing beef producers from all parts of the country how I tackle the task of making money out of suckled calf production. The keen interest which breeders have shown in my ideas and methods prompted me to write this book.

Profitable suckled calf production requires firm positive answers to two important questions:

- Can the capital outlay required to set up a suckler herd be serviced and justified?

- Is there sufficient commitment and ability to make the project successful?

Preface to the Second Edition

Since this book was first published in 1988, the beef industry has undergone radical changes and its integrity has been challenged by various health scares, notably BSE and E. coli. To re-build consumer confidence, farm assurance schemes have been developed by leading retailers which have been based on traceability systems to track the full history of individual animals.

The industry is now highly regulated, but controls imposed by the European Union pay scant regard to technical innovation. Support schemes do the industry a disservice by inhibiting the development of low-cost finishing systems. Infringement of the rules can lead to severe penalties.

Undermining the efficiency of our beef herds is the insidious invasion of the Holstein. The Friesian-cross beef suckler cow has been the traditional, efficient mainstay of the industry. The Holstein influence has wrecked that position with its higher feed requirement, reduced fertility and reduced longevity and poorer progeny grading, all having negative effects on herd performance. Alternative methods of breeding replacements need to be researched and developed. This is a major challenge facing the industry. On a positive note, the delivery of Best Linear Unbiased Prediction (BLUP) by the Meat and Livestock Commission (MLC) to the beef industry has enormous potential to improve efficiency. Those pedigree breeders who use the system to improve the genetic merit of their breeding stock will gain the market lead for supplying high beef-value bulls. BLUP will also play a vital role in organising crossbreeding programmes to produce replacements for suckler herds. As beef competes with alternative meats it is essential that we understand how to build into these programmes a high level of eating quality. It is also essential that we capitalise on the benefits of the hybrid vigour generated by crossbreeding.

Despite the recent changes the basic principles of suckled calf production have changed little, but there are now opportunities to refine production systems to higher levels of efficiency. The requirements demanded by retailers and consumers on farm assurance, traceability and animal welfare issues will have to be met. However the greatest challenge to confront the industry is the drive to reduce the unit cost of production. To do this successfully genetic improvement will play a major role in identifying those animals with improved efficiency.

The suckler systems which have been developed by producers have varying management implications, particularly with regard to capital and labour requirements. Choosing the right system to fit the farm resources and then managing it in the most effective way are vital components for success.

This book is based on my practical experience gained managing three suckler herds over 24 years. I hope that it will provide information and ideas so that the correct choice of suckler system is made and the herd is then managed in the most effective way.

Richard Fuller

June 1998

Abbreviations

ADAS   Agricultural Development and Advisory Service
BLUP   Best Linear Unbiased Prediction
CP   Crude Protein
DM   Dry Matter
EBV   Estimated Breeding Value
EU   European Union
MAFF   Ministry of Agriculture, Fisheries and Food
ME   Metabolisable Energy
MLC   Meat and Livestock Commission

Chapter 1

Principles of Production

Making a profit from suckler cows hinges on the ability to organise each phase of the production cycle into an integrated and efficient system. Before looking at each phase in detail it is important to have a clear understanding of the principles of suckled calf production. These are essentially the same whichever system is adopted to suit the resources of the farm.

Running a successful suckler herd depends on organising each one of three phases objectively and consistently.

1. Breeding (crossbreeding)
 a) Requires a workable plan.
 b) Needs to be based on sound genetic and economic principles.
 c) Choose the optimum cow breeds.
 d) Select high-performance terminal sires.
 e) Select breeds to improve eating quality.

2. Management
 a) Rear the maximum number of calves in relation to cows put to the bull.
 b) Calves must grow efficiently to weaning and must have the ability to finish at the optimum carcase weight, conformation and fat classifications.
 c) Use grassland efficiently in relation to stocking rates and fertiliser usage.
 d) The economic use of feed inputs, especially concentrates, is essential.

3. Marketing
 a) Identify target customers and research their needs.
 b) Produce what you can market, not market what you can produce.
 c) Satisfy customer farm assurance requirements, e.g. traceability and animal welfare standards.
 d) Aim to market finished cattle to well-established retailers.
 e) Look for niche marketing opportunities, e.g. organic beef or the Certified Angus Scheme.

Good quality suckler cow with her one month old bull calf, born in September.

The same cow and calf nine months later. The cow (485 kg) is due to calve again within the next two months; her calf, about to be weaned, now weighs 480 kg.

Thriving February born calves. A compact calving pattern ensures the production of even groups of calves and easy management of cows and calves.

A group of finished suckled bulls showing target weight and conformation.

Cross section of rib showing high muscle content and leanness required to attract premium values. Carcase merit is strongly influenced by the choice of sire.

The calves produced from suckler cows must be regarded as a "crop" and, as with any other farm crop, output and returns should be maximised. A high level of performance from the herd is the result of both appropriate breeding policies and effective management and marketing strategies.

Breeding Policy

For many years the dairy industry has supplied large numbers of beef-cross heifers as replacements for suckler herds. The heavily-built Friesian, when crossed with our native beef breeds, produced highly fertile and efficient cows. They had the ability to maintain body condition on minimal feeding regimes. They were robust and their progeny finished well to high carcase specifications. The gradual expansion of Holstein blood lines throughout dairy herds has produced dramatic changes to the type of animal now being used in suckler herds. These changes compromise efficiency and add to costs. Currently dairy cow numbers are in steady decline and inevitably that will mean there will be reduced numbers of heifers available as replacements for suckler herds.

Disadvantages of Holstein-bred replacements

1. High maintenance requirement.
2. Reduced fertility.
3. Reduced longevity.
4. Poorer progeny grading.
5. Excessive milk yields.
6. Reducing availability.

However, because of the simplicity of sourcing dairy-bred replacements, there will be continued demand for what is, in effect, a by-product of high-yielding, large-framed and poor conformation Holsteins.

On the other hand, it is more difficult to organise breeding programmes to produce efficient beef cows, i.e. low input - high output. The ideal cow type should be of medium size and should exhibit a high level of reproductive ability while retaining good beefing quality and adequate milk production. It is essential that these traits are consistently repeatable to maintain high levels of output.

In addition, optimum working cow weight should be balanced on the one hand to minimise maintenance costs and on the other to produce progeny that falls within the specifications required by meat processors. Current requirements would suggest that the optimum working weight of cows should be 530 to 550 kg. The choice of the most efficient breeds of cows with the highest production potential has a significant effect on the carcase quality and the weight of the calf crop. Performance of the suckler herd can be enhanced considerably by a crossbreeding programme.

Larger cows bred by using heavy continental bulls require higher levels of feeding than do smaller cows and therefore cost more to keep. In general terms, big cows are less efficient. Although they tend to produce heavier calves, the extra calf performance does not compensate for the higher maintenance of their dams. After weaning, the calves from these dams will, if not finished intensively, grow to produce carcases which are too heavy for UK processors. From a practical point of view, stocking rates are reduced, both at grass and in yards. This can be counter-productive when it is important to maximise EU support premium payments in relation to eligible stocking densities.

Hybrid vigour

The most cost-effective strategy to increase suckler cow production is to use crossbreeding to generate hybrid vigour (Figure 1.1).

Crossbred cows perform more efficiently than do pure bred cows. It is well known that the hybrid vigour generated by crossbreeding achieves significant advantages in terms of improved fertility, ease of calving, durability of the dam, and increased milk production which leads to better calf performance. It is important to recognise that the effect of hybrid vigour is enhanced in the calf by breeding crossbred cows to a third breed of sire.

Figure 1.1   **Characteristics of suckler cows and bulls which contribute towards maximising the calf crop at Givendale**

Crossbred dam   **Charolais sire**

Low maintenance requirement   High growth rate
Adequate milk production   High carcase yield
High level of fertility   Lean carcase

Cross generates hybrid   Selected for easy
vigour in cow   calving

Cross gives hybrid
vigour in calf

Herd output maximised

All commercial females should be crossbreeds. Extensive trials in the USA have proved that crossbred cows and calves perform better than do pure-breds. Hybrid vigour improves those traits with low heritability such as fertility, calving ease, calf survival, milking ability, cow longevity and early sexual maturity. All these traits contribute to efficient production. This has been clearly proved in the USA where extensive trials, on a herd basis, have indicated a 25% increase in total calf weaning weight, above the average merit of the pure-bred parents.

When crossbred cows are mated to a third breed then the effect of hybrid vigour encourages enhanced growth rate in the progeny.

The effect of hybrid vigour can be split into three phases. The US trials have shown the following advantages of crossbred cows over pure ones.

Advantages of crossbred cows

Number of cows weaned	+ 7 %
Weight of calves	+ 5 %

Effect of third breed terminal sire

Calves weaned	- 4 %
Weaning weight	+ 8 %

Advantages of crossbred calves

Number weaned	+ 3 %
Weight at weaning	+ 6 %

Total advantage due to hybrid vigour: + 25 %

These results confirm the success of the UK system of using F1 (first cross) dairy bred suckler cows and the use of a third breed terminal sire, e.g. Angus cross Holstein cow crossed with a Charolais bull.

The advantages of hybrid vigour as part of this breeding process may be taken for granted by breeders without them realising the true effect on herd output. Any future attempts to breed replacements must be based on a sustainable breeding system which maximises the effects of hybrid vigour.

Having established the importance of cross-breeding we can now look at the options open to breeders when considering how to organise a replacement policy.

Firstly it is important to recognise that large, highly muscled and lean continental breeds such as the Charolais and Limousin have poor maternal characteristics. Extensive breed analysis carried out in the USA clearly shows that these breeds have poor milk production, long gestation with increased calving difficulty, late sexual maturity and poor fertility traits. They are also large and therefore require high levels of maintenance. However, these breeds are well suited for use as terminal sires because of their superiority in terms of growth rate, meat yield and lean to fat ratio. On the other hand, breeds such as the Simmental, Salers and Gelbvieh confer good maternal traits.

Secondly, it is important to be aware that animals strong in maternal traits are relatively weaker in terms of growth rate and meat yield. That means for breeders who wish to produce their own replacements there are penalties when selling the brothers of the heifers they intend to keep for breeding. However, to improve the conformation and growth of the males they should be left entire. It is therefore important to be aware of the terminal sire line/dam line approach to breeding.

Selection by Estimated Breeding Values

We are now fortunate to have Best Linear Unbiased Prediction (BLUP) analysis in place for all the major beef breeds, both for males and females. This is a system whereby important commercial traits are measured and recorded to produce Estimated Breeding Values (EBVs) for each trait. I will cover BLUP evaluation in more detail in Chapter 5, but for the purpose of selecting females we need to look at the 200-day EBV for milk production.

The 200-day milk EBV for a bull is calculated using data collected from the 200-day growth records of the bull's grandchildren. If the bull is mated to average cows and his daughters to average bulls then we would expect similar 200-day growth figures in all his progeny. But we are likely to see differences and these are expressed in a plus or minus figure, depending on whether the 200-day growth figure is higher or lower.

The importance of 200-day milk

For the pedigree breeder it is important to keep a close eye on milking ability. Pure beef breeds tend to be poor milkers and as selection to increase meat yield and to drive down fat gathers momentum then we will see continued reduction in milk production.

There is a strong case for developing maternal EBVs especially for our native breeds and perhaps also for the Simmental. These should put less emphasis on increasing muscle and reducing fat and more on milking ability, calving ease and eating quality.

When it comes to crossbreeding to produce efficient replacements, 200-day milk EBVs differ in their importance depending on the status of the females involved.

The basic rule is that the further away from Holstein breeding the more important are 200-day milk EBVs.

If, however, a bull is selected as a terminal sire only, then 200-day milk EBVs have no relevance.

Methods of breeding replacements

1. Cross-breds from dairy herds

The national dairy herd will continue to supply replacements and these can be improved by breeders working closely with dairy farmers to select specific bulls to breed an improved type of heifer. EBVs can now identify suitable bulls for this purpose. We should look for bulls with a positive value for calving ease, a plus value for fat and average values for growth and muscle score.

There are several advantages to organising replacements from one source:

    a) Planned entry into the herd.
    b) Control of genetics.
    c) Reduced disease risk.
    d) Uniformity of cow herd.

However, as discussed earlier, it must be remembered that the Holstein is far from the ideal breed on which to base a suckler cow.

2. The half-bred system.

Assuming that most herds start with a dairy-cross type of cow, a proportion of these can be mated to a maternal sire of a third breed, e.g. Limousin cross Holstein cow mated to an Angus bull. To ensure this system is sustainable, a number of dairy-bred heifers need to be purchased annually. Let us assume a 100 cow herd, where the following structure would operate.

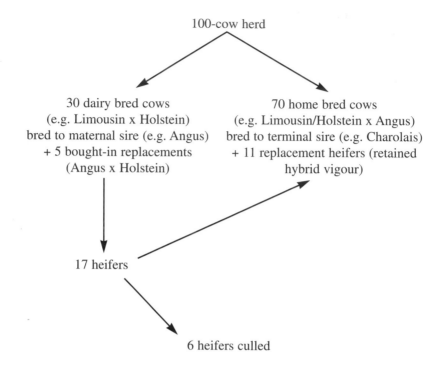

The advantages of this system are:

    a) Control of genetics.
    b) Reduced disease risk.
    c) Uniformity of cow-type is secured.
    d) Hybrid vigour is maintained.
    e) Two-thirds of the herd produce better-muscled calves.
    f) In-breeding by the maternal sire would not occur.

The disadvantages of this system are:

    a) Male calves by the maternal sire would have a lower value than those by the terminal sire.
    b) The system would consist of separate groups and different cow types which would complicate management.

3. The rotational crossing system

This system involves using two or three beef breeds in rotation. It is important when selecting the breeds to identify the strengths and weaknesses in each breed and to use the crossing programme to blend them together to improve performance. It is also important to choose the breeds to suit the climate and individual management styles. For example, the Angus contributes its moderate size, durability and eating quality and its calves can be sold in the UK for a premium under the Certified Angus Scheme. The South Devon contributes to milking ability, carcase yield and low-maintenance cost.

Starting from dairy-bred cows a third sire-breed should be used to initiate the process followed by the use of a fourth sire-breed.

e.g.:   Limousin cross Holstein heifers mated to South Devon.

        South Devon cross heifers mated to Angus.

        Angus cross heifers mated to South Devon.

This pattern of crossbreeding continues with alternate matings of the two chosen maternal sires.

To produce the required number of heifers, approximately 40% of the herd would need to be involved in the rotational crossing programme, the remaining 60% of the cows would be mated to a high performance terminal sire breed, such as the Charolais.

The advantages of rotational breeding:

      a) Hybrid vigour is maintained.
      b) Minimises disease risk through closed herd.
      c) Control of genetics.
      d) 60% of the herd is producing better-muscled calves.

The disadvantages of rotational breeding:

      a) To avoid in-breeding maternal bulls need to be changed regularly. Alternatively AI could be used.
      b) Management of several different breeding groups is complicated, e.g. control of optimum feeding strategies.
      c) There would be considerable variation in cow type and calf performance between generations within the breeding population.

A group of Limousin cross Holstein heifers with calves sired by a South Devon bull. The heifer calves will be retained as replacements for the herd.

Limousin cross Holstein cross South Devon 7 months old heifers. The first step in rotational cross breeding. These heifers will be served with an Aberdeen Angus bull to calve at 2 years old.

Findings at the Clay Meat and Animal Research Centre in Nebraska, USA, have shown that fluctuations in breed composition between generations in rotational cross-breeding systems are likely to cause considerable variation in levels of performance among cows and calves for major traits, unless the breeds used in the rotation are similar in performance characteristics.

4. Composite breeding

Composite breeding techniques have been pioneered at the Clay Centre. The purpose of the research was to develop cow types which can be managed in a similar way to pure breeds but which retain the advantages of hybrid vigour. The researchers have proved that populations of composite cows provide an alternative to more complex systems of crossbreeding, while retaining 75% of the hybrid vigour generated by the first cross.

Their results have shown that composite breeding offers a procedure that is more effective than continuous crossbreeding for utilising genetic differences among breeds to achieve and maintain optimum performance levels for economic traits, on a continuing basis.

Composite breeds are created by crossing pure breeds, which are selected for specific traits. At the Clay Centre 25 pure beef breeds have been ranked for four biological criteria.

      1. Growth rate and mature size.
      2. Lean to fat ratio.
      3. Age at puberty.
      4. Milk production.

The breeds with high values for each trait are then selected and crossed, usually to create a four-breed composite. Their programmes are based on British breeds, such as the South Devon, Black and Red Angus and continental breeds such as, the Simmental and Gelbvieh. The Gelbvieh contributes to early puberty while the Angus improves eating quality by fixing marbling. The Simmental and South Devon contribute to growth, carcase yield and milk production.

The following composite breed is one that was developed as the MARC 11 at the Clay Centre and has proved successful in large-scale commercial production for the Leachman Cattle Company in Montana.

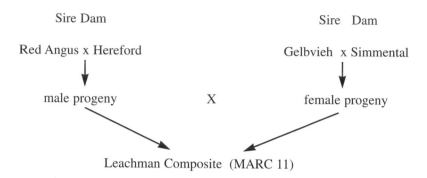

Sire Dam   Sire   Dam

Red Angus x Hereford   Gelbvieh x Simmental

male progeny   X   female progeny

Leachman Composite (MARC 11)

Once the composite breed is formed, the maintenance of at least four family lines are required to avoid in-breeding. Pure bred bulls must never again be used in the programme. New generations of composites can be bred from improved pure breeds and they can then be added into the programme to maintain hybrid vigour.

The difficulty in introducing a composite programme into the UK is the relatively small size of pure-bred herds. This could be overcome by breeders cooperating to form breeding groups. However the traditional independent culture that dominates the beef sector may prove a handicap. However, a well-organised programme would be a valuable objective to improve the efficiency of the UK beef industry.

Composites in the UK could be designed by using existing cross-breeds as well as by deliberately crossing appropriate beef breeds.

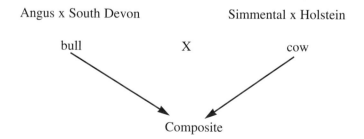

Angus x South Devon   Simmental x Holstein

bull   X   cow

Composite

Composite bred heifers in the USA at 14 months. Showing a high level of consistency. About to be served to calve at 2 years old.

South Devon bull running with pure Angus cows, to produce cross bred heifers for commercial suckler production.

The inclusion of a quarter Holstein would contribute to milk production and early puberty. The Angus would contribute to meat quality and reduced maintenance costs, while the South Devon and Simmental would maintain milking ability and enhance growth rate and carcase yield. To improve the performance of the slaughter generation from the composite cows they would need to be mated to a terminal sire breed, such as the Charolais.

Advantages of the composite system:

a) Maintains the benefits of hybrid vigour without the need for complicated rotational crossbreeding.
b) Creates uniform cow types.
c) Slaughter generations perform to predictable growth and carcase targets.
d) The opportunity exists to build into the composite the optimum economic maternal traits.

Disadvantages of the composite system:

a) Relatively large numbers of pure bred parents are required, from which to select basic stock.
b) If the programme is not well organised then in-breeding can be a problem, so reducing the influence of hybrid vigour.

Management Strategies

Two key features of profitable suckled calf production are that every cow should produce a calf each year, and the growth rates of the calves should be maximised. Both of these targets must be borne in mind when considering which management practices to adopt.

Figure 1.2 shows management practices which play a part in the process of maximising the calf crop from the Givendale herds.

Understanding the nutritional requirements of the cow at different stages of the production cycle and feeding her correctly at each stage ensures that she both produces a healthy calf and is in calf again within 3 months. Growth rates of calves in the suckler herd are maximised firstly by ensuring that the cows are fed as required to produce adequate amounts of milk, and secondly by offering the calves creep feed. At Givendale, the basis of the efficient management of both cows and calves to maximise calf output is the achievement of a compact calving pattern.

Figure 1.2　　　**Management practices to maximise the calf crop**

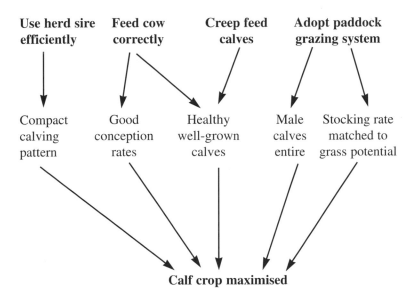

Profitability is also increased by paying close attention to the cost of feeding the herd and identifying areas where savings can be made without adversely affecting cow or calf performance. Finally, the adoption of a paddock grazing system not only allows the efficient use of grass, but also enables male calves to be kept entire, thus increasing their potential growth rate.

Compact Calving Pattern

As already noted, a compact calving pattern has a dominant role to play in the success of a suckler herd. A calving spread of not more than 9 weeks should be the aim, although an even shorter period would be advantageous. It is crucial, however, to get the suckler cow back in calf within 12 weeks of calving, as failure results in disruption to herd management which in turn leads to lost production.

Suckled Calf Production

The key to achieving a compact pattern of calving is knowing the right level of feeding and the correct body condition for the cow at the critical stages of the reproductive cycle, as discussed in Chapter 3. A second important factor which also influences the calving pattern is the efficient use and performance of the herd sire. This aspect of suckler herd management is considered in more detail in Chapter 5.

Figure 1.3 illustrates the main advantages of a compact calving pattern, all of which contribute to maximising herd output.

Figure 1.3   **Advantages of a compact calving pattern**

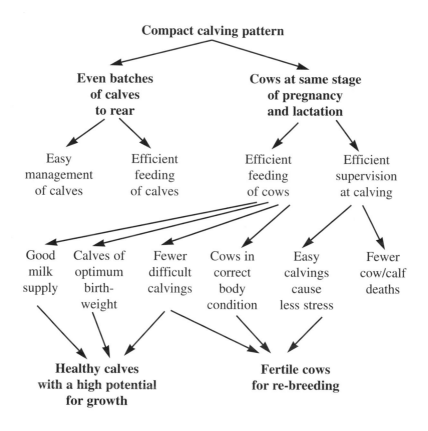

Chapter 2

Systems of Production

Systems of suckled calf production are usually first identified by the time of year when the cows calve. The most common systems involve suckler cows calving at grass in the autumn, in buildings during the winter, or at grass in late spring (May and June). In a few cases summer calving (i.e. calving in July and August) is practical, but production tends to be poor in terms of output per cow.

Autumn calving has become less popular due to the high costs of feeding lactating cows during the winter period. In many cases this system has been abandoned with preference given to grass-based production centred around spring calving.

The season of calving has a number of implications for the organisation of the herd. Therefore, when planning a suckler system the first question which must be asked is "When would be the best time to calve the cows?" Serious thought must be given to this question, and to arrive at the correct answer the resources of each individual farm must be examined.

The first part of this chapter highlights some important factors which influence the choice of calving season. In the second part, the organisational implications of autumn, winter and spring calving systems are considered.

Choosing the Calving Season

The main criteria that influence the decision when to calve are:

- Feeding
- Housing
- Labour
- Calf disposal

Out-wintered Luing cows on the Isle of Islay.

Feeding outwintered cows in a permanent trough using a concrete standing area. The cows are allowed a 'loaf' on a sacrifice area of moorland.

Feeding

As we shall see later, the majority of management decisions concerning the suckler herd are taken with feeding in mind because the costs associated with feeding have a major impact on performance, and therefore profitability.

Silage is probably the most common form of forage fed to suckler cows, and the relatively high costs which are associated with silage-making mean that full consideration must be given to using this feed economically. Silage may be derived from the early cutting of permanent grass which is part of the grazing area later in the season, or from specific temporary leys grown as a break crop in an arable rotation. The latter option has the disadvantage of taking land out of the arable cropping programme, thereby increasing the cost of the silage because the ley has to carry its share of the arable overheads. Advantages, however, are that a ley produces heavy cuts of grass and it also provides a 'clean' grazing area for weaned calves in the autumn.

When considering suckler cows as part of a lowland arable system, the opportunity arises to use arable by-products such as stock feed potatoes, or even to grow a specific high-yielding crop such as fodder beet. The most obvious arable 'waste' product, however, is straw of which thousands of tonnes are produced each year.

It has been shown that untreated wheat straw, supplemented by a small amount of cereal, is an adequate feed to over-winter May or June calving cows. If the wheat straw is treated with ammonia the resulting increase in digestibility ("D"-value) also enables lactating cows to be fed successfully on a straw-based diet, with cereal supplementation. Straw, either treated or untreated, therefore has a major role in suckler cow production as a relatively cheap form of forage in most situations, although transport costs generally make it a more expensive commodity on hill farms.

Housing

Winter housing of suckler cows is desirable in the majority of cases because damage to land is avoided, and because housed animals require considerably less energy than do out-wintered ones. There may be a case, however, for out-wintering spring calving cows on the milder west side of the country when well-drained land is available.

Suckler cows can be housed in a wide range of buildings. The availability of straw for bedding is an important factor when considering the type of housing, but it must be stressed that expensive and elaborate buildings cannot be justified in the majority of cases. The principle of housing suckler cows is to give them plenty of air and a dry sheltered lying area.

Open-fronted pole barn sheds are ideal, but if straw is scarce and expensive then cubicles or cow kennels are suitable, provided the cows do not have to calve in them. Slurry disposal may be a problem with these latter types of housing which use little straw.

When planning winter accommodation it is important to remember that ideally groups should consist of no more than about 40 cows in one yard. There are several reasons for this:

(i)   If the cows are to be mated during the winter period then 40 cows are enough for one bull to cope with.

(ii)   Calves, when housed with their dams, tend to perform better in smaller groups and creep feeding is easier to regulate.

(iii) With larger groups of cows and calves there will inevitably be a percentage of animals that succumb to environmental stress, which will result in poor performance.

It is also important to be able to separate the breeding herd into groups depending on the nutritional needs of the cows. Ideally the herd should be split into three groups:

(i)   Replacement heifers.
(ii)   Second calvers and thin cows.
(iii) Mature cows.

Building design and feeding systems should allow the economic use of labour and should be sympathetic towards the well-being of the livestock. Stressful conditions, such as badly ventilated buildings or insufficient feeding space, will contribute towards depressed performance. Buildings which create labour-intensive situations may lead to jobs not being done properly, to lack of attention to detail and to inadequate observation in a rushed situation.

Labour

On arable farms a danger is that the suckler herd is treated as a minor part of the business because it is there to utilise an area of rough grazing or some parkland. The herd may well not be big enough to justify a full-time stockman, and the lack of specific responsibility may lead to management practices being delayed or not carried out at all.

Cubicle house for suckler cows.

Open fronted housing creates a healthy environment for suckler cows.

This has the effect of creating a downward spiral of poor performance which wastes farm resources and can result in the business actually making a loss. It is therefore important to appreciate the need to identify the peak labour requirements of the suckler herd and to plan accordingly.

Supervision during the calving period demands the most concentrated labour input of the year. Therefore it is advisable to calve the cows during a relatively slack period on the farm. For instance, it would be unwise to calve during the harvest period on an arable farm unless there was labour specifically available for the suckler herd. In the upland situation it may be inadvisable to allow calving to coincide with the lambing period, because performance of both operations could suffer.

It is also important to recognise the need to spend extra time with the herd during the mating period to ensure that the bull is working properly. Assuming that all is well at such a critical time is not good enough, and may lead to catastrophic losses if a bull fails to serve the cows or if his fertility drops.

Calf disposal

It is desirable to maximise the output from each cow in the herd, and in general the older the calf when it is sold, the higher its value. Autumn-born calves secure the highest prices when sold the following autumn. Calves born in the same year as they are weaned are lighter, so to maximise the return per cow the calves need to be housed during the following winter for a growing-on or finishing period. In a lowland situation, therefore, where arable by-products and adequate buildings are available, selling finished cattle rather than weaned calves should be considered.

On the other hand, in a hill situation forage may be scarce and expensive, and there is a danger that the cows compete for resources with the weaned calves. This can result in neither the cows nor the calves being fed adequately, and the sale of weaned calves before winter may then be preferred.

These examples illustrate that careful planning and budgeting is necessary to identify the best method of calf disposal to suit the resources of individual farms.

Autumn Calving

Autumn calving in September, October or November is a high input/high output system. Because the cows are calving towards the end of the grazing season, they tend to be fairly fit, and possibly over-fat, so supervision during the calving period is demanding.

Once housed, the herd has to be fed well, not only to cater for lactation, but also to ensure that good conception rates are achieved. Stocking density in the buildings is lower than for spring calvers to allow extra space for the growing calves and for their creep area. The labour requirement for an autumn calving herd during autumn and winter is relatively high, although this situation may be advantageous if spare labour from the arable enterprise can be utilised during the winter.

At the end of the winter the cows and calves may be turned out to grass together, but stocking rates should make allowance for the area required by the fairly large calves. Alternatively the calves may be weaned and then either turned out on to good 'clean' grassland, or kept inside for a finishing period. In this latter situation the cows may be turned away on to inferior grazing and stocked tightly until they calve again.

Weaning the calves at turnout particularly suits those farms where an area of rough hill is available for the cows. Generally, however, in lowland situations the herd is best turned out in its entirety in order that, in addition to grazing good quality spring grass, the calves can benefit from the flush of milk from the cows. The high output of the system is realised when heavy calves are weaned in early August.

Winter Calving

Winter calving in buildings during January, February and March falls between autumn and spring calving in terms of input and output. On some farms the main drawback to the system is an increased risk of disease, particularly scouring, in young calves - but with careful management and the use of modern vaccines the rearing period should be relatively trouble-free.

A September born calf looks for its first suckle of colostrum.

Winter calving herd. New born calf in a well bedded straw yard.

February calving cows and 3 months old calves with a high Beef Value bull.

New-born twins. Cows that produce twins and suckle twins need to be separated from the rest of the herd. This will allow for extra feeding to meet the increased nutritional needs of the cow and will ensure adequate milk for the calves.

Winter calving systems have the advantage of allowing the turnout of fairly strong calves which are able to take full advantage of a flush of milk from their dams. The cows can be fed more cheaply through the winter than can autumn calving animals, and they should be stocked more tightly at grass after turnout because they have smaller calves. The calves, when weaned in October or November, are 3 months younger than autumn born calves, and they weigh approximately 65 kg less, but because of lower feed costs and higher stocking rates returns from the two systems are similar. After weaning the calves may be sold or, alternatively, if buildings and feed are available, retained and finished. This latter option ensures maximum output from the herd.

Spring Calving

Spring calving in April, May or June is a low input/low output system which may suit farms where large quantities of straw are available. The dry cows can be fed relatively cheaply during the winter, either on ammonia-treated straw alone, or on untreated straw plus a small amount of supplement. The cows calve outside, having lost weight during the winter so calving difficulties are minimised. After calving, the availability of grass which is of a high feeding value means that the cows conceive successfully as their body condition improves.

The biggest drawback to this system is that calves are still relatively small by the time they are weaned in the autumn. To make the system economically viable, after weaning the calves must be housed either for finishing rapidly or for feeding-on as stores, which can then be sold as and when resources dictate.

Chapter 3

Management of the Suckler Cow

The function of the suckler cow is to produce a good quality calf which, within the adopted production system, weighs as much as possible in relation to her own body size. This should be achieved as economically as possible every 365 days.

Three areas of suckler cow management are particularly important: nutrition, health and calving. In addition, condition scoring and pregnancy diagnosis are useful management tools.

Nutrition

Adequate nutrition of suckler cows has a profound effect on reproductive performance and profitability of the herd. The majority of management decisions are related to feeding, and feed costs make up the highest proportion of the costs of suckled calf production. In addition serious problems, illustrated in Figure 3.1, arise when cows are underfed.

Strategies to feed a cow adequately and economically depend on her size, her body condition, the stage of her reproductive cycle and whether she is still growing. In order to ensure good performance and the cost-effective use of feed, it is important to have a clear understanding of the nutritional needs of the cow during the various stages of her reproductive cycle.

Nutrition in relation to the reproductive cycle

When we study the job that the suckler cow has to do, we see that she has to maintain her own body, feed her calf, come into heat, conceive promptly and then deliver a live calf. The range of nutritional requirements for the cow to do this job well is considerable, and it is important to recognise that the nutritional resources available to her are used in a specific order. The priority of a suckling cow is to feed her calf by producing milk. Her next priority is to maintain herself and, finally, reproduction comes last on her list. So we can see that if a cow does not get adequate nourishment, then reproductive performance is the first to suffer.

Figure 3.1   Consequences of the undernutrition of suckler cows

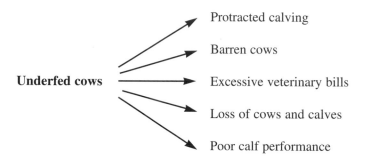

The most critical period in the reproductive cycle is the time between calving and mating. The energy requirement of the cow is at its highest then because she is trying to put on weight after calving, the calf's demand for milk is at its peak and she is trying to get her reproductive system ready for re-breeding. Adequate intakes of relatively high quality feeds are needed at this time, and it is also important to continue to keep energy intake at a high level for 6 to 8 weeks after mating to ensure the successful implantation of the foetus.

In mid-gestation energy requirements are much lower and cows can afford to lose body weight with no detriment to themselves, their calf at foot or their foetus. During this stage it is important to be aware that savings in feed costs may be possible.

In late gestation cows must not be allowed to get too thin because 80 percent of foetal growth takes place during the last 8 weeks of pregnancy and, in addition, the cow is preparing herself for lactation. Breeders who worry about difficult calvings tend to run too much condition off their cows, which is a bad policy because it is difficult and expensive to feed thin cows for lactation as well as for the necessary improvement in condition which is required for successful re-breeding. Rather than regain body condition, cows may well channel the extra feed provided after calving into excess milk production which can result in scouring calves, which in turn lead to management problems.

By contrast, the opposite situation is that cows should not be excessively fat at calving because calving difficulties are more likely due both to fat deposits in the pelvis and also to an increase in the weight of the calf. Emphasis, however, is placed on the term 'excessive', especially where heavy continental sires are used.

Feeding for high conception rates

Mating at grass
Cows that are mated while they are grazing are ideally suited to achieve high conception rates, provided that an adequate allowance of good quality grass is available and intakes are not restricted for any reason. In addition, cows should not have been allowed to get too thin prior to calving.

Mating in winter
The level of feeding for cows that are mated in the winter is of critical importance. To achieve the best conception results the herd must be housed in 3 separate groups, which require different levels of feeding:

(i) **First calved heifers** require extra feed in a non-competitive environment to allow continued growth as well as lactation.

(ii) **Second calvers and thin cows** also need extra feed. Second calvers are notorious for refusing to gain weight after calving. As thin cows are almost invariably the older ones, the heavy milkers or suckling twins, the need for extra feed is obvious.

(iii) **Mature cows in good condition** are better housed away from the other two groups to prevent bullying and general environmental stress. If it is not practical to house the herd in 3 groups then the first 2 groups should be housed together, away from the mature cows. Savings in feed costs can be made with the mature cow group.

It is important to recognise the need to feed cows on a rising plane of nutrition up to the mating period to allow for the increasing milk demand of the calves. It is also important to feed high phosphorus minerals to the cows because phosphorus has been shown to aid fertility. Finally it is not sound practice to change dramatically the ration of cows during the mating period because sudden changes in diet can lead to reduced conception rates.

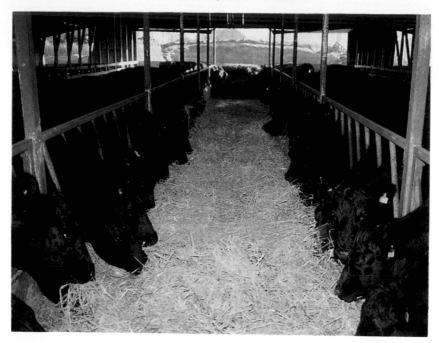

Winter calving cows eating ammonia treated wheat straw in a cubicle shed.

Figure 3.2   Condition scoring

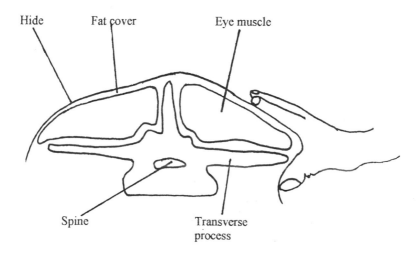

Condition Scoring

Condition scoring should be used to assess the nutritional needs of cows at the critical stages of their reproductive cycle. Initially scoring should be done by handling the cows individually, but when sufficient experience has been gained it should be possible to score them by eye.

When handling the cows the procedure is to grip the loin halfway between the last rib and the hip bone. The hand is placed on top of the loin with the fingers pointing towards the spine, and the thumb is pressed round the transverse processes of the spine to feel the fat cover over the tips of the processes, as shown in Figure 3.2.

Body condition is scored on a six point scale from 0 (very thin) to 5 (very fat) as follows:

Condition score	Description of loin
0	The spine stands out and the ends of the processes feel very sharp.
1	The spine is still prominent but the transverse processes feel less sharp.
2	The spine is less prominent and the transverse processes feel cushioned by the increased fat cover.
3	The transverse processes can only just be felt.
4	The transverse processes cannot be felt.
5	The transverse processes cannot be felt under a thick layer of fat.

Target condition scores

Table 3.1 shows the target condition scores for autumn and spring calving cows at different stages of the reproductive cycle.

The dangers of allowing cows to become too thin or too fat prior to calving has already been noted, and identifying cow condition with a scale of figures allows a better definition of the correct body condition.

Condition scoring. Note position of hand over loin area.

Table 3.1   Target condition scores

Stage of reproductive cycle	Target conditioning score	
	Autumn calving cows	**Spring calving cows**
At calving	3	2.5
At mating	2.5	2
Mid-pregnancy	2	3

Autumn calving cows

Calving cows in the autumn at a condition score of 3 reduces the need for expensive feeds after calving and allows half a score drop in condition up to the bulling period. If the cows were to be calved at a condition score of 2, large amounts of expensive feed would have to be given to provide sufficient nutrients for lactation in addition to allowing an increase in body weight. After mating and successful implantation of the embryo, cow condition can be allowed to drop to a score of 2 in mid-pregnancy, enabling savings to be made in feed.

Condition score: rear view of cows in, from left to right, score 0, score 2 and score 4.

After turn out, which occurs during mid-pregnancy, body condition should increase to score 3 during the grazing season. A higher score than 3 could lead to calving problems, especially in herds where heavy continental sires are used. Very tight stocking of the cows after weaning may be necessary to stop the cows putting on too much condition prior to calving, especially in a good year for grass growth.

Spring calving cows

As we have seen, the advantages of spring calving over autumn calving are lower winter feed requirements and the utilisation of nutritious grass to achieve good conception rates. The target score for calving is 2.5 and, because a high intake of nutrients from spring grass results in a rising plane of nutrition, cows should conceive at condition score 2. During the grazing season the condition of cows should gradually improve to score 3, which allows subsequent savings in feed due to the use of body reserves through the winter, prior to calving at score 2.5.

Advantages of condition scoring

Condition scoring helps to:

- Maximise conception rates.
- Allow economic use of feeds.
- Identify thin cows which can then be given preferential treatment.
- Avoid calving problems.

Herd Health

Certain diseases and conditions are of particular importance where suckler cows are concerned and it is worthwhile discussing at this stage the most common problems. Suckler cows do not enjoy the individual attention that dairy cows receive twice a day as they go through the milking parlour, which means that it is important to be constantly observant for diseases such as mastitis. An alert stockman can do much to minimise the effects of disease.

Cows taking magnesium syrup, supplied at turnout to grass to prevent hypomagnesaemia.

Stress is a prime catalyst for triggering many conditions, so stressful situations such as rough handling or badly ventilated buildings should be avoided. When a disease can be controlled by adopting preventive measures, they should be used because prevention is more economical and humane in the long term than treatment or the loss of animals. It must also be remembered that drugs should not be relied upon to compensate for bad management practices - they should only be regarded as an aid to sound stockmanship.

Common diseases and disorders

Hypomagnesaemia (staggers)

This condition, which is due to inadequate magnesium in the blood stream of the cow, is a killer. Lactating cows or cows whose calves have just been weaned are at the greatest risk. Hypomagnesaemia occurs most frequently when cows are turned out onto lush spring grass, but it may also occur when cows graze a flush of grass in the autumn. Stress associated with weaning or moving cows, sudden drops in temperature and poor weather conditions all seem to be catalysts for the development of this disorder.

Prevention is by far the best course of action to take because the first symptom is often a dead cow. The feeding of magnesium-rich supplements during the high risk periods helps to alleviate the problem. When a cow suffering from hypomagnesaemia is found alive she must be treated very quickly with an intravenous injection of magnesium salts

Milk fever (hypocalcaemia)

Milk fever, which occurs in cows before, during or more commonly after calving, is caused by a fall in calcium and possibly phosphorus levels in the bloodstream. It is not possible to take preventative measures, so swift action is necessary to treat affected animals. This takes the form of an intravenous or sub-cutaneous injection of a mixture of calcium and phosphorus.

Older cows which milk well are most susceptible, and animals which have had an attack are liable to a repetition when they next calve. The incidence of the disease appears to increase when cows calve outside in cold, wet weather, and in these cases hypomagnesaemia may also be involved to some extent

Mastitis

This is a very nasty condition and in the worst cases cows can lose their milking ability, abort or even die. The treatment of mastitis can be extremely time consuming and is not always successful. Quarters can still be lost after several weeks of stripping out the udder and giving antibiotic therapy.

The symptoms to be on the look out for in suckler cows are:

- Clots hanging from the end of a teat.
- Swollen quarters - special note should be taken of cracked teats which are too sore for the cow to allow her calf to suckle from, as this situation often leads to mastitis.
- A cow which kicks at her calf when it tries to suckle may have a painful, infected udder.
- Lameness in the back leg may be caused by the pain of an infected udder.
- A cow which stands away from the rest of the herd and is obviously ill must be suspected of having mastitis.

There are many different types of bacteria which cause mastitis; some are more virulent than others. Flies may be responsible for spreading certain types of infection, such as summer mastitis, from cow to cow.

The following preventative measures can be taken, although they are not foolproof:

- Drying cows off after weaning is a critical time in the control of mastitis. The udder should be treated with a long acting 'dry cow' antibiotic and the teats should be washed and dipped with a proprietary teat dip. The cows should also be sprayed with an anti-fly preparation, which can be done using a knapsack sprayer.
- Housing the cows on straw after weaning for 3 to 4 days helps to stop milk production, so reducing the risk of an attack of mastitis.
- Pour-on anti-fly preparations are very useful to keep flies at bay where they are a problem, such as in wooded areas or near water. These products are particularly useful in protecting maiden heifers against summer mastitis.

Infertility

The inability to breed regularly causes infinitely greater financial losses than those for all other diseases put together. The immediate loss is not so apparent as it is with a specific disease, but delay or failure to breed interrupts production and therefore reduces profit. There are many causes of infertility but the most common ones are under the control of the stockman. They include:

- Poor feeding and body condition. As explained earlier, cows must be in good condition at mating. They must also be given an adequate supply of minerals, which must be high in phosphorus, especially during the mating period. Copper also has an effect on fertility and deficiencies should be rectified on the advice of a veterinary surgeon.

- Infection of the uterus. This can occur after a cow has retained the afterbirth or when attention to cleanliness at calving has been inadequate.

- Infection of the ovaries. This generally manifests itself as irregular or absent heat periods.

Mating records are very useful to identify cows which have a breeding problem. If cows do not hold to 2 services or if they are not seen bulling at all then they should be examined by a veterinary surgeon. Appropriate action can then be taken.

Administration of a copper bolus to correct copper deficiency.

Parasites

Parasitic infections interfere with the performance of suckler cows and should therefore be controlled. The efficiency of modern drugs makes control simple and economic.

Worms should be treated by using a broad spectrum anthelmintic on suckler cows shortly after housing in the autumn. This treatment rids cows of all types of worms and also kills hibernating larvae, so preventing pasture contamination the following spring. Due to the fact that modern anthelmintic are so efficient, once-a-year treatment should be adequate to prevent a build-up of all types of worms in the herd. Cows or heifers which are ôbought-inö should always be wormed as soon as they come onto the farm to avoid contamination of pasture and a breakdown in the system.

Lice are common in untreated cattle during the winter months. There are several preparations on the market which control this parasite and they should be applied soon after cattle are housed in the autumn. However, infestation may re-occur by late February, in which case the cows should be dusted with louse powder to eliminate itching and hair loss.

Pregnancy Diagnosis

The confirmation that, following mating, cows are pregnant is useful for several reasons. In particular it is an early warning guide for:

• Planning herd replacements
• Efficient cull cow marketing
• Identifying a sub-fertile bull
• Identifying fertility problems in cows

In spring calving herds, cows should be pregnancy tested shortly after weaning in the autumn. Barren cows can then be identified and sold. Autumn calving cows are best examined just before turn out.

When spring and autumn herds are run on the same farm, non-pregnant cows can be moved from one herd to the other. Pregnancy testing is necessary in this situation to ensure that non-pregnant cows can be re-bulled at the correct time. Only younger cows should be considered for this procedure, and then only after a satisfactory internal examination by a veterinary surgeon. Older cows and those with recurrent fertility problems should be culled.

Calving

The profitability of the suckler herd depends on the cows giving birth to live calves, and every effort should be made to ensure that this happens. The target must be for every cow to have a live calf. The majority of cows calve quite normally without assistance, but it is never possible to predict which cow is likely to need assistance, so constant observation is necessary during the calving period.

If calving occurs outside, every effort should be made to make the stockman's job as easy as possible. The calving field should be as close to his home as possible and it should be accessible with a vehicle. Night-time patrols are more conveniently carried out with a vehicle, which should be equipped with good lights. If cows have to be calved a long way from the farm buildings then it is well worth constructing a catching pen in the field.

Once calving is in full swing the herd should be checked every 3 to 4 hours around the clock. The majority of calves are lost between midnight and 6am because the 3am check is the one which is most often missed.

Emphasis should be put on achieving and maintaining a tight calving period, with a target of 85 percent of cows calving in the first 4 weeks. Quite clearly this situation helps to make monitoring the herd easier.

The breed of bull used on the herd has an effect on calving problems, as shown in Table 3.2, but problems can arise with all breeds. If the herd is managed properly, there is no reason why calf mortality should be any higher in a herd where a Charolais bull is used than in one where an Angus bull is the chosen sire. There is certainly more skill and attention to detail required with a Charolais bull, but this is well rewarded by the extra value of the calf crop.

Stages in calving

A thorough understanding of the physical act of calving is very helpful in achieving good results in practice. The stages in calving can be split theoretically into four, but it must be recognised that calving is basically a continuous process.

1. Preliminary stage

This usually starts several days before calving takes place. The vulva becomes swollen and the udder increases in size and becomes tender. The ligaments on each side of the tail head slowly slacken until calving is imminent, when they can no longer be felt.

Wooden catching pen for use when assistance with calving is required. This pen is placed with access from 3 fields

Calving at night in the wooden catching pen.

Table 3.2 **Assisted calvings and calf mortality for different breeds of sire**

Sire breed	Assisted calvings (%)	Calf mortality (%)
Charolais	9.0	4.8
Simmental	8.9	4.2
South Devon	8.7	4.0
Devon	6.4	2.6
Limousin	7.4	3.8
Lincoln Red	6.7	2.0
Sussex	4.5	1.5
Hereford	4.0	1.6
Angus	2.4	1.3

Source: Meat and Livestock Commission

2. Dilation of cervix

When this stage starts the cow becomes uneasy and usually separates herself from the rest of the herd. As dilation progresses the cow assumes the typical stance of arched back and tail held straight out. Milk often drips from her teats and she repeatedly lies down and gets up. The water bag then appears. She becomes more restless, and when the water bag bursts she licks the fluid off the ground and often calls to the calf.

3. Expulsion of the calf

When the cervix is fully dilated the contractions become more intense and more frequent. Each contraction pushes the calf a little further up the pelvic canal until the front feet and then the head are pushed out, followed by the rest of the calf. The sooner the cow starts to lick her calf the better because the licking process stimulates and dries the calf.

4. Expulsion of the afterbirth

This usually happens within 1 or 2 hours of calving. Sometimes the afterbirth is retained, however, and it has to be pulled away after about 4 days. As this can lead to infection of the uterus, veterinary advice should be sought.

Calving problems

Calving is not always straightforward, and knowing when and how to assist is critical in saving the lives of cows and calves. The first step that all stockmen should take is to acquire a calving aid and then learn to use it correctly. This simple, inexpensive piece of equipment enables one man to calve cows single-handed in situations where 3 or 4 men would normally be needed. It has a positive, controllable action which is ideal for precise assistance.

A normally-presented calf is born with both front feet appearing first, closely followed by the head. Any departure from this position results in an assisted calving. There are several different ways in which a calf can be incorrectly presented and examination of the cow can identify the problem. Readjustment must then be carried out before delivery of the calf can take place safely.

Examination of the cow should always be carried out with care and meticulous attention to cleanliness. The hands and arms should be washed thoroughly in a suitable disinfectant and soaped well, together with the vulva of the cow. This procedure helps to prevent genital infections which could lead to subsequent breeding problems.

The use of a calving aid for a normally presented calf.

(a)

(b)

(c)

(d)

Calving problems: posterior presentation of calf. Note the use of a calving aid. Immediately after the calf is delivered, it is held up by the hind legs to drain fluid from the mouth, throat and lungs (d).

Signs of trouble

If a cow stands away from the rest of the herd and shows early signs of calving but nothing appears after about two hours, she should be examined. The chances are that there is something wrong. Older cows seem to know if the calf is lying incorrectly for normal presentation and they will give up trying to calve. This can also happen when twins are tangled up with each other. In situations like these, calving the cow should proceed with some urgency, as a calf left for too long inside the uterus will die from lack of oxygen.

If the front feet have appeared and the cow has been pushing hard for about an hour, but she is obviously making no progress, then she should be caught and given assistance. The head of the calf is probably stuck in the pelvic canal and this should be checked. If it is possible to pass the hand between the head of the calf and the pelvis of the cow, strong positive traction should be sufficient to pull the calf free. If the head locks tight veterinary assistance should be summoned.

The correct time to apply traction to the calf is when the cow strains with each contraction. The main advantage of a calving aid is that it keeps tension on the calf's legs between contractions, so preventing the calf from slipping back into the cow.

If the back legs of the calf appear first then its hooves point with the soles uppermost and the cow always needs rapid assistance. Strong traction is usually necessary to deliver the calf. As soon as the calf is born it is advisable to hang it up by its hind legs so that any fluid in the air passages can drain out.

Action when calving problems arise

Expertise in calving can only be acquired through experience built up over a period of time, and complicated calvings are best dealt with by a veterinary surgeon until sufficient experience has been gained. It must be remembered that once a cow has started to calve then she should not be left any longer than an hour before she is examined. The longer a cow is allowed to exert useless contractions on a jammed calf, the less chance there is of the calf surviving and the greater is the chance of the cow damaging herself.

Chapter 4

Culling Strategy and Herd Replacements

The consistent and long term success of suckled calf production depends on the regular replacement of breeding stock. A specific culling and replacement policy is therefore needed.

Culling

The basic principle of culling is to remove those cows from the herd which for various reasons are having a detrimental effect on the profit potential of the business.

Reasons for culling

Barren cows

Failure of cows to become pregnant is probably the most common reason for culling. The main cause is usually inadequate feeding prior to and during the mating period, and therefore most culling of barren cows could be avoided. However, every year there are likely to be a few repeat breeders for non-nutritional reasons, and these have to be replaced.

Udder problems

These include udders which have been damaged after severe attacks of mastitis. In older cows pendulous udders can result in calves not being able to suckle properly.

Age

It is advisable, as a general rule, to cull cows by about 11 years of age, after they have had 8 calves. As cows get older their milking ability declines, which results in poorer calf performance. Their reproductive performance also declines, which can result in difficult calvings and protracted calving periods, and they are more prone to problems such as milk fever. Another reason for culling cows is that it is generally more difficult to get old cows into good condition for the cull market.

Performance

The performance of suckler cows varies from individual to individual, just as it does in dairy cows. Regular recording of milk production in the dairy herd can identify individual cow performance. In a suckler herd, however, the only means of assessing milking ability is by recording the weight of the calves at weaning. Cows that regularly produce calves below the average calf weaning weight for no obvious reason should be culled.

Before the BSE crisis cull cows were a valuable source of income in UK herds and it was well worthwhile feeding the cull cows to increase their sale value. However, the introduction of the Over Thirty Month Scheme (OTMS) in 1996 fixed the price paid by the government for each animal going into the scheme. This was at a level well below the pre-BSE values and the shortfall has had a serious impact on the income of suckler producers. Providing a cow weighs the minimum of 550 kilos then maximum compensation is paid. When the OTMS is eventually abandoned it will be worth marketing cull cows at maximum weights to increase income.

Herd Replacements

In the average situation the target annual replacement rate for a suckler herd should be about 16 per cent. This gives cows an average herd life of 7 years.

Source of animals

The final decision on how to source replacements depends on the adopted breeding policy. This was discussed in Chapter 1.

The following options are open for a herd replacement policy. They can be separated into three categories of females.

1) Bought-in dairy-bred: a) heifer calves
                               b) yearling / bulling heifers
                               c) cows - dry or suckling

2) Bought-in suckler-bred: a) weaned heifer calves
                               b) yearling / bulling heifers
                               c) cows - dry or suckling

3) Retained suckler-bred

Bought-in replacements

The introduction of outside stock into a breeding herd is a risky business because of the challenge that it presents to the health status of the herd. However, these risks can be minimised when replacements are bought at an early age and mixed with the cows from the herd so that they have the opportunity to acquire a level of herd immunity before they calve. Another disadvantage of buying in replacements can be the unknown genetic value of the animals. This problem can be resolved by co-operating with other farmers to organise specific breeding programmes to produce stock with high maternal merit.

Buying dairy-bred heifer calves

This is the cheapest way of obtaining herd replacements. In many respects, it is the most reliable, especially if calves can be bought from the farm of origin, where specific bulls can be used to produce progeny with good maternal traits. About 20 per cent more calves should be purchased than are required to allow for the disposal of undesirable types, non-breeders and casualties. A potential limitation of this method of obtaining herd replacements is the lack of grazing and feeding facilities for young heifers within the farm system.

Buying yearlings or bulling heifers

This is a good method because the desired type of heifer can be selected and control can be exerted over the choice of bull which is used on them. Heifers can also be mated to fit in with the calving season of the herd. About 10 per cent extra heifers should be bought to allow for non-breeders and casualties.

Buying in-calf heifers or cows

This is usually a more expensive method of obtaining herd replacements which leads to faster calf production, but there is less control over the time of calving and the type of bull used.

Buying calved heifers

This is again an expensive option but at least the risk is taken out of calving the heifers. However, there is a danger of bringing disease into the herd especially when cattle are bought at a market. Again there is little control over the type of bull used to sire the calves.

Buying calved cows

Unless the source of stock is known then this method has the least to recommend it due to the uncertain history of this class of animal. The introduction of disease to the herd is a greater risk than with calved heifers and unless absolutely necessary it is a risk which is not worth taking.

Retaining home-bred heifers

If organised well, then this method will ensure the integrity of the health status of the herd. However, as discussed in Chapter 1, it is vital to breed heifers with optimum maternal traits, while maintaining hybrid vigour. In a single-season calving herd it is vital to achieve first calving by the age of two years. To secure this target attention has to be paid to organising a management strategy which produces heifers big enough to bull at 15 months of age. This will be achieved by selecting heifers born early in the calving season and by designing a feeding regime which grows heifers to a target bulling weight of approximately 380 kilos.

Selection of heifers

Selection of breeding heifers is best done before they are mated, as those which are not considered to be good enough can either be sold as stores or they can be finished for slaughter.

Selection of breeding heifers should be based on:

- Sound conformation
- Frame size
- Good legs and feet
- Steady temperament

Stocky, fat heifers are undesirable and should therefore be culled.

Management of herd replacements

The management of replacement heifers must ensure that they achieve target bulling weights. These can vary depending on how the animals are bred, but generally heifers should weigh between 360 and 400 kg when mated to bulls selected for easy calving. EBVs play an important part in this selection process, as described in Chapter 1. There is a strong case for calving heifers at two years of age to ensure that they come into production at the optimum time to minimise prolonged rearing costs.

Angus x Friesian heifer calves; potential herd replacements.

A group of cross-bred replacement heifers running with a Limousin bull which was selected with a positive Calving Value to ensure easy calving.

It is important to recognise the need to breed from well-grown heifers, because calving smaller animals increases calving problems and may retard their growth rate and eventually reduce their mature weight. On the other hand, it is extravagant to allow heifers to get too big before they are mated because the investment in the animals does not bring returns soon enough. Young heifers must also not be overfed because their breeding and milking ability can be affected adversely by excessive fatness.

Autumn calving heifers. Calving takes place 6 weeks before the main cow herd to give the heifers more time to recover from parturition.

It is sound management practice to serve heifers so that they start to calve four weeks before the main herd. This allows more time for recovery after calving and for regular oestrus to get underway. Heifers generally have small calves, so calving early in the season allows more time for them to reach better weights by the time they are weaned. The choice of the bull that is used on heifers must be based on calving ability because heifers are prone to calving difficulties, the incidence of which should be minimised as far as possible. The use of native breeds is the safest course to take, but of the continental breeds the Limousin and Blonde d'Aquitaine are probably the safest breeds of sire to use.

After calving, in a housed situation heifers should be separated from the rest of the herd to allow better control of feeding. They require extra feed in a non-competitive environment to allow for growth as well as for lactation. It may well be necessary to wean the heifers before the cows, especially in a poor grass year, in order to give preferential feeding to improve their condition before they calve again.

It is important to remember that heifers must be looked after with care during their first lactation, to ensure that they are in good enough condition for rebreeding to take place successfully. Failure to achieve this will either waste valuable time and resources or may result in the heifer having to be culled, with the subsequent loss of a potentially productive animal.

Chapter 5

The Herd Sire

The herd sire has a major effect on the profitability of suckled calf production. The old saying " the bull is half the herd" is an understatement of the real influence that a bull exerts on herd performance. In general, the range of influence of individual beef bulls within each herd can be almost as great as the range between breeds. Therefore the choice of individual bulls within each breed is of paramount importance. The dominant traits that a bull passes on to his progeny, such as growth rate and carcase yield, have high levels of heritability. The choice of sire also influences calving ease, but to a lesser extent, as this trait has a low level of heritability.

Choice of Sire Breed

The choice of the breed of sire to be used on a herd depends on several criteria, including calving difficulties, breed of dam and calf disposal.

Calving difficulties

The heavier the sire breed the more likely the incidence of calving difficulties, and calf mortality rates are likely to increase. The choice of breed should therefore reflect the level of commitment to monitor closely the herd during the calving period. At the same time, however, calving difficulties can be considerably reduced by correct cow management and by selecting individual bulls which are likely to give fewer calving difficulties.

Supervision during calving is desirable in all herds to ensure the maximum number of calves born alive. But particular care taken where, say, a Charolais bull is used, is very well rewarded by the extra value of the calf crop.

Breed of dam

The need to use bulls of a different breed to that of the cow has already been stressed (Chapter 1).

Calf disposal

The heavy breeds of cattle, such as the Charolais and Simmental, produce progeny that finish for slaughter at higher weights; they also provide leaner carcases than do our native breeds. Consequently, if the calves are sold as stores they command premium prices, and if they are retained for finishing, the added value of the carcases is of great benefit to the farm business. The effect of sire breed on calf weaning weight is shown in Table 5.1.

Selection of Individual Bulls

Selection of the individual bull must be taken very seriously because the final choice has a profound effect on the profitability of the herd. There are great variations of type and size within each breed, so it is advisable to learn something of the proven bloodlines of the chosen breed to help in the final selection. Sons of cattle which have outstanding breeding records behind them are obviously desirable.

Table 5.1　　　　**Effect of sire breed on calf weaning weight**

	Type of herd		
	Lowland	*Upland*	*Hill*
Sire breed	*200-day weight (kg)*		
Hereford	208	194	184
	Difference from Hereford cross (kg)		
Charolais	+32	+33	+21
Simmental	+24	+28	+14
South Devon	+23	+27	+16
North Devon	+17	+21	+7
Lincoln Red	+14	+20	+5
Sussex	+7	+13	+2
Limousin	+7	+10	+2
Aberdeen Angus	-14	-12	-8

Source: Meat and Livestock Commission

Givendale Josh, bred at Givendale. Josh is a highly ranked bull for growth and muscle traits and his Beef Value places him in the top 1% of the breed.

BLUP Evaluation (Best Linear Unbiased Prediction)

The delivery of BLUP to the beef industry by the UK Meat and Livestock Commission was the most significant development that has ever been introduced and its successful implementation has enormous potential to improve efficiency.

BLUP technology is a sophisticated technique which was developed by highly-skilled teams, both at the Roslin Institute and the Scottish Agricultural College. The programmes they have developed are world-leading in their ability to identify those traits in beef cattle which are directly associated with improving producers' margins. They do this by identifying those sires whose progeny grow more efficiently and are of the correct conformation to satisfy the carcase specifications of meat plants.

The BLUP programme uses the performance data of pedigree beef cattle to calculate Estimated Breeding Values (EBVs). This data is collected by the Signet Beefbreeder National Beef Evaluation Service and the National BLUP Runs are the responsibility of the MLC.
BLUP separates genetic factors influencing animal performance from non-genetic factors. It then produces a series of Estimated Breeding Values (EBVs) which are estimates of the worth of the animal's genes for the

recorded traits. The genetic factors are those heritable traits which are passed on by parent stock to their offspring via their genes. The non-genetic factors are those affecting animal performance through different feeding regimes and environmental conditions. The influence of environment and genes on performance are evaluated simultaneously in an one-stage process ensuring that all the available records are combined. In order to achieve this it is essential that genetic links exist between contemporary groups. This is achieved by individual bulls having offspring in as many herds (contemporary groups) as possible. This genetic linkage between herds then allows BLUP to use the performance data of related offspring, which are expected to perform with greater similarity than unrelated animals.

In the UK the calculations are carried out using an Individual Animal Model, which is the most sophisticated method of BLUP analysis available. Performance records on individual animals for each major trait are included in the evaluation as well as all the available records from relatives in different pedigree herds.   Information on all the recorded traits is analysed simultaneously using multi-trait BLUP.   This is an accurate method of analysing the data because information on one trait helps to establish the animal's breeding values in other genetically correlated traits, e.g. a high value for 200-day growth is likely to result in a high value for 400-day growth. On the other hand when desired traits are negatively correlated then multi-trait analysis makes it difficult for an animal to have favourable EBVs to match the breeders aims, e.g. 400-day growth and high muscle scores are usually negatively correlated with calving ease; however, when an animal occurs to match the former with easy calving then it is highly likely that it is favourable for all the traits.

Estimated Breeding Values

The BLUP programme produces a series of EBVs which are measured in the same units as the traits themselves. They are expressed relative to a common base-line year for all animals of the same breed in all the contemporary groups, e.g. a bull with an EBV for 400-day growth of +50 kg is genetically 50 kg superior for that trait when measured against the "average" bull born, of that breed, in the base year. A bull passes on half his genes to his progeny, therefore the calves sired by this bull would be expected to weigh 25 kg heavier at 400 days when compared with calves sired by an average bull born in the base year.

Establishing a common base year enables direct comparisons to be made across genetically linked breeding populations. This is important in the UK where small herd sizes would otherwise hinder the rate of genetic improvement. Across-herd EBVs increase the opportunity to select superior breeding animals from an enlarged genetically-linked base. They can then be used to monitor genetic progress year by year and any movements in breeding values will confirm the effectiveness of breeding strategies.

Accuracies of EBVs

The accuracies of EBVs are recorded in percentage units and they are presented along side the EBVs. They indicate how close the EBVs are to the true breeding values of an animal and they provide a guide as to the likelihood of an EBV changing as more information on the animal becomes available, e.g. a high accuracy indicates that the EBV is robust and is not likely to change much when additional performance data of relatives is added to the calculations.

Several factors affect the accuracy of EBVs:

1. The amount of recorded performance information on the animal and its relatives.
2. The degree of heritability of each trait.
3. The strength of the correlation between traits.
4. The number of contemporaries recorded - the more the better.

The Beef Value

EBVs are presented for each measured trait. This allows breeders to decide how much emphasis to place on them. Matings can then be planned to compensate for weaknesses to improve the overall standard of breeding populations in pure breeds. The individual performance trait EBVs are combined into a multi-trait selection index, which is presented as the Beef Value. This ranks animals on their estimated overall genetic merit for growth and carcase traits. It is calculated using economic weightings which are relevant to current market conditions to ensure that each trait is given the appropriate amount of emphasis in the index. This is the most efficient method of improving several traits at once in a pedigree breeding programme. It is also the best guide for commercial breeders when selecting terminal sires.

The beef value is designed to improve returns by focussing on driving forward carcase weight, leanness and conformation scores. The following measurements are included in the multi-trait calculations:

- 200-day growth (kg)
- 400-day growth (kg)
- muscle depth (mm)
- fat depth (mm)
- muscle score (points)

It is possible to use the EBVs for these traits to predict overall genetic merit, because they are in direct correlation with the selection traits. For example:

1)   High 400-day growth EBVs indicate heavy carcase weights and a linked increase in Beef Value.

2)   High muscle score and muscle depth EBVs indicate higher carcase yield, so will increase Beef Value.

3)   Negative back fat EBVs indicate leaner carcases, so increasing Beef Value.

The Calving Value

Three EBVs are combined to make up this multi-trait selection index:

1) gestation length (days)
2) calving ease (score)
3) birth weight (kg)

The aim of the Calving Value is to improve returns through the genetic selection of calving ease and gestation length, while taking into account birth weight. It is a useful sire selection tool because it highlights the effect of using a particular sire on the ease with which his progeny are likely to be born. A bull with a positive Calving Value is likely to sire calves which are born more easily than calves sired by a bull with a negative value.

The advantages of two selection values

In general, bulls which perform well in terms of growth and conformation tend to be heavier at birth with the increased risk of difficult calving. Conversely, bulls whose progeny calve easily tend not to perform as well. The two values , therefore, give breeders flexible opportunities to select bulls for specific production systems. However, a minority of exceptional bulls have positive correlations between ease of calving and calf growth rate and therefore satisfy both criteria. These are clearly a valuable source of genetic material.

The selection of bulls with high Beef Values will leave progeny that show improved performance when compared with average bulls for the following characteristics:

1) heavier carcase weights
2) higher meat yield - better carcase confirmation
3) leaner carcase

The selection of bulls with high Calving Values will result in progeny which have the following advantages over average performance:

1) shorter gestation lengths
2) born more easily

Having studied all the relevant information which is available, the final selection must be made with the physical soundness of the bull in mind.

Desirable physical features include:

• Sound legs and feet, combined with good action.
• Clean shoulders and a good, level topline.
• Well-fleshed and good muscling over the rump.
• Minimum brisket and belly.
• Good length.
• Masculine head and appearance.
• Steady temperament.

Undesirable points which should be avoided are:

• Straight or sickle-hocked hind legs.
• Bulls with small testicles, as these can lead to fertility problems.
• Young bulls which carry a lot of fat or belly because they have probably been overfed and will quickly 'melt' when put on a commercial breeding regime. Fat bulls are not a sound long term investment.

Management of Bulls

Considerable capital is invested in a bull which has the desired characteristics, and to get the best out of this investment he must be managed with care and attention. If a bull is looked after properly he can still serve cows successfully when he is 7 or 8 years old. However, the average working life of a bull is much shorter than this, which in many cases is due to inadequate care of the animal.

It is important to keep the herd sire in good physical condition, not only to ensure a long working life to justify his purchase price, but also because the condition of a bull prior to a working session has a direct bearing on the conception rate that he is likely to achieve. He must be physically strong but not fat, his feet must be in good condition, he must be fertile and he must have the desire and the ability to deposit his semen in the cow.

Nutrition

The correct feeding of the young bull is very important to ensure his continued growth and proper development. A bull is usually purchased when he is 15 to 18 months of age, and if the right bull has been selected then he will still be growing at about 1.5 kg per day. This growth rate should be maintained by feeding him about 9 kg a day of a mineralised cereal and sugar beet pulp ration of about 14 per cent crude protein. The growth rate and quantity of feed suggested refer to large continental breeds; smaller British beef breeds will have a lower growth rate and will require less feed.

The age at which bulls mature depends on the breed, with bulls of larger breeds maturing later than bulls of smaller breeds. As a general rule, the larger breeds grow rapidly until they are about 2.5 years of age but their final mature size is not reached until they are about 4 years old. Bulls over 2 5 years old require less cereal-based feed because they are able to utilise predominantly forage-based diets to sustain a declining growth rate.

Exercise

In addition to adequate nutrition, exercise is important when the bull is laid off work. A small, well-fenced paddock is useful for this purpose during the summer, and in the winter a bull pen with sufficient exercise room should be provided. Before constructing a bull pen, advice on the legal requirements concerning bull penning and handling should be sought from the Health and Safety Executive.

Foot care

The majority of bulls have sound feet which require only minor trimming periodically, but unfortunately there are some bulls which need frequent attention to ensure their continued mobility. Correct foot care prolongs the working life of any animal, but foot trimming is a particularly neglected task where bulls are concerned because they are difficult to handle without the right facilities. All too often nothing is done until the bull goes lame, which often occurs when he is working, possibly due to the extra strain that he exerts on his feet when mounting the cows. The bull may then have to be rested to allow time for the injured foot to heal, and a serious financial loss may be incurred because another bull may have to be hired or bought in a hurry, in order to get the cows in calf on time.

Foot injuries can be caused by foreign objects penetrating the sole of the hoof. I have seen, to my horror, both bulls and cows in fields and farmyards which are littered with scrap iron and old posts with nails sticking out of them. It is amazing that the owners seem to be unaware of the dangers that these objects pose to their valuable livestock. A deep, penetrating wound into a hoof can cause permanent damage and the eventual untimely slaughter of an animal, invariably accompanied by veterinary costs.

Routine foot trimming should be considered essential, but the length of time between treatments depends on how quickly the feet of an individual bull deteriorate. Generally speaking, all bulls should have their feet checked once a year, but more frequent inspections may well be necessary for bulls with a history of bad feet.

There are two ways of approaching foot trimming. Firstly, the bull can be sedated by a veterinary surgeon who then trims his feet while he is laid out. Secondly, a foot trimming crush can either be bought or made on the farm and the job can then be tackled by a competent stockman who has received adequate training. Generally speaking, foot trimming is not a difficult task and it is a skill which is well worth learning because routine work can be undertaken regularly without the worry of escalating veterinary costs.

Foot trimming a bull in a tip-over crate.

Assessment of fertility

It is of course essential for a bull in a commercial situation, serving a lot of cows quickly, to have good sperm production. A good guide to the likely fertility of a bull is the size and shape of his testicles. They should be well developed, feel firm and be of equal size.

Mating records should be kept in order to verify a bull's conception rate. It is also advisable to check by observation that a bull continues to serve his cows properly because it is not unknown for a bull to damage his penis in such a way as to make serving impossible.

As a bull gets older the quality of his semen tends to deteriorate, and it is therefore well worth having it checked annually once a bull reaches about 6 years old. This can be done on the farm by a veterinary surgeon with the appropriate equipment.

Working a young bull

Young bulls must not be expected to get a lot of cows in calf when they are at an early age. Age at sexual maturity varies depending on the breed and it can also vary between individuals of the same breed. Generally speaking, the bigger the breed, the later sexual maturity is reached.

Collecting semen from a bull to determine his fertility status.

Initially a young bull should be mated with a cow which is in standing heat, away from other stock. Once he has been seen to serve the cow then he can be turned into a herd of no more than 20 cows. It is a bad policy to run a young bull with a larger number of cows because poor conception rates could result, giving a protracted calving period. Larger numbers of cows could also result in the bull losing too much condition and even suffering growth retardation.

It is important to keep a careful watch on a young bull to make sure that he continues to work properly. It is also advisable to take him away from the herd after 3 weeks and to replace him with an old bull. This strategy covers for the young bull in case he is sub-fertile or infertile, and keeping a check on the cows which return for re-mating provides valuable information on his success.

If the first working session of a young bull occurs during the winter, special care must be taken to ensure that he is adequately fed and it may be necessary to feed him individually. In a confined area bullying from older cows can be a problem. There is a danger of chills developing after sweating. To prevent injury to the bull it is important to ensure adequate bedding and to avoid slippery surfaces such as those found in cubicle sheds.

Chapter 6

Managing the Suckled Calf

For a calf to achieve its full potential it must have adequate milk from its mother. It must also have access to creep feed at certain times of its life. The health status of a calf also has a considerable bearing on how quickly it grows and on the accumulation of labour and veterinary costs. Considerable benefits both in terms of growth and conformation can be gained by leaving the bull calves entire, although adjustments to management practices may be necessary.

Nutrition

Colostrum

The future well-being of a calf depends on the early absorption of sufficient quantities of colostrum. A calf is born without any antibodies to protect it from bacterial infections such as septicaemia, joint ill, pneumonia or scour. The antibodies which are acquired by the calf for this protection are carried in the colostrum, which is also high in nutrients and vitamins.

The ability of the calf to absorb antibodies decreases after 6 hours from birth, so it is essential to observe that a newly born calf has suckled its mother within this period. Special care should be taken with first-calved heifers to make sure that they mother their calves and that they actually allow their calves to suckle. Where a calf is unable to suckle its mother because of injury or some other circumstance, the cow must be milked and 1.5 litres of colostrum given to the calf for its first feed. In this situation it is unwise to try to force the calf to drink, so the colostrum should be given with the aid of a stomach tube to avoid waste and to avoid passing fluid into the lungs of the calf.

There are no substitutes for colostrum so it is advisable to have a deep frozen supply of colostrum from a cow which has calved previously, to cover for emergency situations when colostrum cannot be provided by the calf's dam.

Milk

A good supply of milk is essential for calves to continue to grow to their full potential. The advantages of hybrid vigour in this respect have already been discussed (Chapter 1), and the nutritional needs of cows to ensure adequate milk production were covered in Chapter 3.

Creep feeding

Creep feeding has a beneficial effect on growth rate at times when the nutritional value of milk and forages is insufficient to maintain the growth potential of the calves. As calves grow older the effect of milk on growth rate is reduced because, with the development of the rumen, emphasis moves from milk to forage as the major source of nutrients. The creep feeding of calves should therefore be considered in situations when milk yield is reduced due either to a restriction in the amount of feed offered to cows or to a decline in the nutritional value of grass. The most effective way to utilise creep feed is to anticipate occasions when the calves are likely to gain the most benefit, and then to increase intake slowly during these situations.

New born calf suckling colostrum, which is vital for the transfer of antibodies from cow to calf.

Area for creep feeding autumn born calves. The vertical bars are too narrow for cows to enter the area, but they are wide enough apart for calves to pass through to eat the creep feed.

Creep Feeding Strategies

Autumn born calves

The creep feeding of autumn born calves is well worthwhile especially during February, March and April, because milk production from the cows declines during this period. The reduction in milk yield comes about not only because of advancing lactation, but also because savings may be made in the feed of cows at this stage of their reproductive cycle. It is more efficient to feed cereals directly to the calf at this time than it is to feed the cow for lactation.

Autumn born calves start to take small quantities of creep feed by the end of November. As they grow older intake increases, until by the end of February the calves may be eating 2 kg per head per day. By this stage the rumen has developed so it is also beneficial to give the calves good quality silage or hay. Provision of a separate area is necessary where the calves can lie away from the cows and where creep feed and forage can be offered.

When the herd is turned out in the spring, creep feed is not needed because the calves benefit from a good supply of nutritious grass, in addition to a flush of milk from the cows. Later in the grazing season, when the quality of grass and the supply of milk declines, it may be desirable to feed the calves a cereal supplement. This is a time-consuming task because the calves have to be separated from the cows every day to be trough fed, although after an initial training period it is not as difficult as might be imagined.

The advantages of this sort of supplementation are apparent when:

    a.   The calves are housed for finishing because the change in ration is less abrupt, or

    b.   The calves are sold, especially at suckled calf sales, as they have 'bloom' and have gained extra weight.

Winter born calves

There are no advantages in creep feeding calves which are born after January, because the supply of milk from their dams should be adequate to carry them through to turn-out. After the herd is turned out in the spring, the calves are able to take advantage of the boost in milk production from grazed pasture.

Creep feeding spring born calves prior to weaning.

Creep feeding can, however, make a significant contribution towards calf growth from July onwards, and creep feeders should be put out during June so that the calves get used to going into them. In order to ensure that all the calves have access to the ration, one 5-space feeder should be available for every 20 calves.

Initially, small amounts of creep are taken, but as grass growth and milk supply decline during July, the intake of creep feed increases. This pattern continues during the late summer, until the calves are weaned in October. Calves which are fed in this way not only weigh considerably more when they are weaned, but they also take more readily to concentrate feeding when they are housed, so reducing stress and weight loss at this time.

Spring born calves

Calves which are born in May and June benefit from creep feeding in the autumn for the same reasons that have been indicated for winter born calves. The method of feeding should be exactly the same, but because the calves are younger creep feeding need not start until the end of August.

Calf Health

Diseases in suckled calves must be avoided as far as possible to ensure that the calves grow to their full potential. Correct management can go a long way towards minimising the risk of diseases, but when problems occur then it is essential that treatment is carried out swiftly and positively. It is important to anticipate the likely situations which can lead to calves developing diseases so that preventative measures can be taken.

The essential role that colostrum plays in building up resistance to bacterial infections of young calves has already been considered. Stressful situations associated with either handling calves or with adverse environmental conditions can trigger off diseases such as pneumonia or scouring, but the incidence of disease can be reduced by good management, as follows:

- Handling should be kept to a minimum, and routine procedures such as dis-budding and castrating should be done in favourable weather conditions to reduce stress.

- Buildings should be well ventilated and calves should have a dry, draught-free lying area which must be kept well bedded.

Dis-budding a calf in a home-
made calf crush designed to
reduce stress.

- Weather conditions are of particular significance where respiratory diseases are concerned. Spring and winter born calves are vulnerable to adverse weather, so a careful watch should be kept on calves during cold, windy and wet spells. Housed calves in the winter are at risk from outbreaks of pneumonia, which seem to be directly dependent on the prevailing conditions. The higher the atmospheric humidity, the greater the number of airborne pathogens and the greater the risk of an outbreak of pneumonia. It is therefore important to be extra vigilant during foggy conditions or during a thaw after a cold spell.

- One of the risks associated with calving suckler cows in straw-bedded yards is the problem of scouring calves. This situation can largely be overcome by vaccinating the cows before they calve. It is important to remember that scouring calves must be treated very quickly, with veterinary advice. Navel infections and joint-ill can also be a problem if navels are not treated soon after the calf is born.

- Deficiency of certain trace elements such as copper or selenium can be a problem because they are essential nutrients for calves and cows. Veterinary advice should be sought if, for no apparent reason, a proportion of animals in a herd is unthrifty.

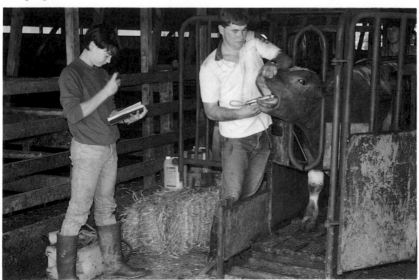

Weighing and worming a calf at weaning.

Copper bolus being given routinely to a calf which will graze copper deficient pastures.

Healthy February born calves. The result of vaccinating the cows to control Rotavirus, hygienic conditions at calving and well bedded yards.

Intravenous saline treatment to correct dehydration in a scouring calf.

Stomach tubing a calf with
electrolytes as part of the
treatment for scour.

Leaving Bull Calves Entire

The fact that bulls perform more efficiently than steers provides an
opportunity to increase the value of male calves. Trials have shown that bulls
grow 15 per cent faster than steers. When bull calves are finished intensively
the growth advantage is enhanced by improved carcase grading and leanness.
Typical targets for this system of finishing are a growth rate of 1.7 kilos per
day and an average final slaughter weight of 650 kilos.

The following points must be considered when bull calves are left entire:

- Cows with bull calves must be split from those with heifer calves when
  the calves are about six months old. This ensures that young bulls cannot
  harass heifers.

- Separate, well-fenced grazing areas must be available for calves older
  than about six months.

- Cows that have twins of mixed sexes must have their bull calf castrated
  so the family group can run with the heifer calf group.

- Cull cows which have bull calves at foot and which have not been put
  back in calf must run with the heifer group, with their calves castrated.
  Cows which come bulling in a group which is suckling bull calves can
  cause considerable disturbance and stress, therefore it is essential for
  cows to be pregnancy-diagnosed in good time so that only pregnant cows
  run with this group.

Chapter 7

Grassland Management and Forage Conservation

It is important to recognise the need to achieve efficient grassland performance, and to plan the grazing and conservation requirements of the herd with care. Stocking rates are significantly influenced by the effect of fertiliser on grass production, and this effect can be exploited to increase the size of the suckler herd. The majority of lowland suckler herds graze marginal land which is unsuitable for cultivation ù land which is either steep, stony, badly drained or parkland. But the need for conserved winter feed must be met, and where silage has been selected as the main forage the utilisation of temporary leys may be necessary. Silage may also be taken from areas of permanent grassland which can then be grazed later in the season.

Permanent Grassland

Managing a permanent grass grazing system is a delicate balancing act between grass supply and cow demand. The supply part of the equation involves encouraging the sward to produce a high output of grass, while on the other hand ensuring that stocking rate is at an adequate level to make full use of the grass produced. Cow demand is largely governed by the size of the animal and the stage of her production cycle.

The main obstacle to achieving the balance is variation in grass production from season to season, due mainly to fluctuations in rainfall and in temperature. Good sward management involves a number of factors and failure in any one will have a detrimental effect on final grass production, as illustrated in Figure 7.1.

Figure 7.1   Factors contributing to a productive sward

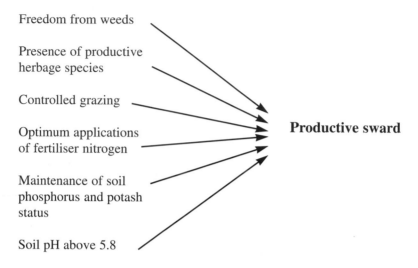

Freedom from weeds

Presence of productive
herbage species

Controlled grazing

Optimum applications
of fertiliser nitrogen

Productive sward

Maintenance of soil
phosphorus and potash
status

Soil pH above 5.8

Freedom from weeds

The sight of grazed pastures which are knee high in numerous weeds all doing their best to choke out the productive grasses in the sward is all too common. It is difficult to understand why farmers allow this situation to occur when they go to great lengths and expense to keep their arable crops free from weeds ù after all, grass is a crop. But traditionally it has been mis-understood, mis-treated and therefore mis-managed.

The removal of weeds is as essential in a grass crop as it is in an arable crop. Several selective herbicides are available but care must be taken not to damage any clover which may be present in the sward.

Sward type

In a situation where grassland is difficult to re-seed, for example on a steep hillside, encouragement of indigenous grasses is probably the best course to take. If, however, the land is ploughable then it may be worthwhile to undertake a re-seeding programme using a perennial ryegrass and white clover mixture to improve the pasture.

Clover can be an important component of the sward, and it has a major contribution to make in suckler cow and sheep enterprises. The 'D' value of clover varies little during the grazing season and trials have shown that liveweight gain and milk yield are improved with a higher intake of clover. Its inconsistency in adverse weather conditions, however, leads to disappointing nitrogen fixation in some seasons.

Controlled grazing

It has been shown recently that sward height has a considerable influence on the utilisation of grass and on animal performance. The established guidelines for grazing swards indicate that the average height of the sward should be maintained between 6 and 10 cm. The shorter height is appropriate in the early part of the grazing season, while taller swards are required in the autumn.

Sward height is used as an indicator of the likely nutritional value of the grass. The most nutritional grass is young, fresh growth, which has a high 'D' value. If a sward is understocked, grass height gets out of control and the cattle graze only parts of the area. In this situation the undergrazed areas soon run to seed and are wasted.

Controlled grazing therefore involves manipulating stocking rate to prevent plants going to seed. This can be achieved by setting up a series of paddocks which are grazed in rotation or by set stocking with the correct number of animals. The latter option is in many ways the better of the two, although it is the most skilful to perfect and it relies on past experience of the potential of the available grassland. Certainly, cattle seem to be more settled and therefore 'do' better in a set-stocked situation. If grass growth does get out of control and seed heads appear, the pasture must be ôtoppedö to encourage fresh growth.

Fertiliser nitrogen

Grass growth in early spring is influenced by temperature. A technique developed in Holland has resulted in the widespread use in the UK of a method, known as the T sum, to decide when to apply the first dressing of nitrogen fertiliser. The T sum is calculated by adding all the positive daily air temperatures from January until the accumulated total of 200°C is reached. This is known as 'T sum 200' and nitrogen should then be applied as soon as conditions allow. In most areas the T sum 200 is reached by mid-March, well before the more traditional April time of first application.

A 4-wheeled drive tractor top-dressing permanent grassland at the end of May.

Surveys conducted by the Agricultural Development and Advisory Service (ADAS) and by the Institute for Grassland and Environmental Research (IGER) have shown that grazing land has an advantage of increased output when nitrogen is first applied at T sum 200.

The total amount of nitrogen applied in individual cases depends on the intended intensity of stocking. A higher stocking rate obviously requires more nitrogen than a lower stocking rate. Past experience of the capability of a particular area of land to grow grass with increasing nitrogen levels can result in a steady rise in stocking rate.

Soil pH and phosphate and potash status

Soil analysis can determine the status of soil acidity (measured as pH), and phosphate (P) and potash (K) levels. For good grass growth the soil pH must be 5.7 or above, and a soil index of 2 for both potash and phosphate are required – any deficiencies should be corrected by liming or applying the correct fertiliser.

Cows and calves grazing well-managed pasture. Controlled grazing maintains grass height at optimum for pasture utilisation and for animal performance.

Buffer feeding the winter calving herd with straw in September.

Buffer Feeding

In situations where stocking rates are high, there is always a possibility that grass will run short in adverse weather conditions such as drought or unseasonably low temperatures. If a grass shortage does occur then it may be necessary to give the herd extra forage-straw, hay or silage, in order to meet the nutritional requirements of the cows and calves. This is termed buffer feeding.

Grass Conservation

The conservation of grass is a vast subject which is covered in many publications. This section therefore is limited to a brief look at the basic principals of silage quality and how different forms of conservation may be utilised in suckled calf production.

In the majority of cases conservation needs are probably best met by the utilisation of temporary leys. Modern Italian ryegrasses are ideal for silage making, giving high yields of material with a good feed value. A temporary ley also makes an ideal arable break crop and is useful for grazing weaned calves after the silage cuts have been taken. In situations where permanent grassland is flat enough for forage equipment to work safely, it may be possible to capitalise on spring grass by taking an early cut of grass on part of the area. Grazing can then be extended onto the conservation area as grass growth declines during mid-summer.

Contrary to the opinions of many, I believe that silage quality is very important where the feeding of suckler cows is concerned. I have already emphasised the relationship between feeding, body condition and reproductive performance (Chapter 3). The success and profitability of herds that calve in the autumn and winter hinge on good quality silage. Spring calving herds on the other hand only require a maintenance level of feeding during the winter so silage with a lower 'D' value and a higher fibre content is adequate - the advantage here is that a bigger crop of material can be taken when the grass is more mature, so increasing stocking rates.

Forage harvesting grass for silage. Good quality material will reduce the need for supplementary feeding.

The silages which I aim to feed to the autumn and winter calving herds have the target analysis shown in Table 7.2.

Table 7.2   Target analysis for silage

	Target
pH	4.1
DM (%)	25 to 30
Ash (% in DM)	9
CP (% in DM)	16
ME (MJ/kg DM)	11.0
'D' value	70
Ammonia N (% total N)	8

Silage of this quality requires minimal cereal supplementation to achieve the desired cow condition for good reproductive performance and for adequate milk production. Poorer quality silage requires additional feeding of cereals to achieve the same results, which inevitably increases costs.

The silage making system at Givendale is based on temporary leys because the permanent grassland is too steep to cut. The grass is cut when young, just before the seed heads appear, leaving a 3 inch stubble to avoid soil contamination. A drum mower/brush conditioner is used and the swath is left as wide as possible to increase wilting rates. The grass is picked up with a precision-chop machine, usually 24 hours later, depending on weather conditions.

Big-bale silage is becoming more popular as bale wrapping techniques become more reliable. On many farms big bale silage has replaced hay, because it is less dependent on dry weather during the summer.

Baled silage does not require expensive silage clamps and is often made by contractors. It is particularly useful where relatively small quantities of silage are made, where there is an opportunity to conserve grass surplus to requirements later in the season, and where transport of silage from a clamp would be difficult or labour-intensive.

Chapter 8

Recording Herd Performance

The Need for Records

The importance of developing well-organised management standards to achieve profitable production has already been emphasised. Recording plays a vital role both in the effective day-to-day supervision and in the longer term planning of a suckler herd.

Records clearly have an important part to play in the preparation of costings. In addition, through the actual process of recording attention is focussed on the performance of the herd, and this stimulates a desire to improve and to refine management techniques.

The main elements which need recording are:

1. *Feed:* accurate records of the type and quantity of concentrate and forage fed.

2. *Grassland:* records of grazing and conservation areas to evaluate grassland utilisation.

3. *Calf performance:* calf growth rate calculated from calf weights recorded at different periods of the year, such as at birth, at turnout, at weaning and at housing.

4. *Breeding:* records of mating and calving details for individual cows, to assess the performance of both dams and sires.

5. *Financial:* records of the purchases and sale of stock and the cost of other items, such as veterinary treatment.

Suckled Calf Production

Table 8.1 **Performance targets for UK suckler herds**

	Continental sire[a]		British sire[b]	
	Autumn calving	Spring calving	Autumn calving	Spring calving
Lowland herds				
Calving period (max. days)	90	80	90	85
Calves weaned/100 cows bulled	93	94	95	95
Calf daily gain (kg)	1.1	1.1	0.9	1.0
Sale weight of calf (kg)	361	297	300	230
Stocking rate (cows/ha)	1.7	1.8	2.0	2.4
Upland herds				
Calving period (max. days)	90	85	90	85
Calves weaned/100 cows bulled	94	92	95	95
Calf daily gain (kg)	1.0	1.1	0.9	1.0
Sale weight of calf (kg)	395	318	300	230
Stocking rate (cows/ha)	1.5	1.6	1.6	2.0
Hill herds				
Calving period (max. days)	100	90	100	90
Calves weaned/100 cows bulled	87	85	90	90
Calf daily gain (kg)	1.0	1.0	0.8	0.9
Sale weight of calf (kg)	360	316	260	220

a e.g. Charolais
b e.g. Hereford

Source: Meat and Livestock Commission

Keeping on Target

The first step is to set production targets. The recording system may then be used to ensure that the targets have been met, so keeping the herd on course for profitable production.

The MLC has gained a wealth of experience of suckler herd management in the UK over many years, and the Beefplan service which it offers tailors production targets to suit individual farms. The average production targets published by the MLC for suckler herds in 1986 are shown in Table 8.1.

Herd Records

Herd records play a very important part in the day-to-day running of the herd. Cow and calf identification forms the basis for recording essential information which can then be used when making management decisions.

Individual identification is important for:

1. Breeding records.
2. Resolving health problems.
3. Determining the age of cows.
4. Matching cows with their own calves.

Every herd in the country has its own number, approved by the UK Ministry of Agriculture, Fisheries and Food, which is stamped on ear tags which must be on two tags, one in each ear of every calf born in the herd. The first step in setting up an individual recording system is to implement, in addition, a cow numbering index. This is best achieved with the use of numbered plastic ear tags. A record must be kept of both the herd tag number and the cow tag number, as plastic tags may be accidentally torn out of the ear, leaving the herd tag as the only means of identification. The age of cows can be expressed by using a letter prefix on the plastic tag which relates to the year of entry into the herd. Cows can then be easily dated at a glance, which proves useful when making management decisions.

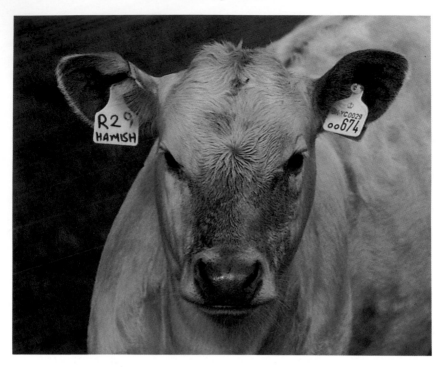

The tags on this calf comply with MAFF regulations. The plastic primary (left ear) and the secondary metal tag (top of right ear) carry the herd number and individual number. In addition the plastic management tag (right ear) carries the number of the dam and the name of the calf's sire.

The ear tags of the whole herd should be checked twice a year and any that are found to be missing or illegible must be replaced with the same number. It is essential to carry out one of these checks prior to calving because every calf should be tagged to correspond with the number of its dam.

Breeding Records

As we have seen, increased production from a suckler herd can be achieved by selecting breeding stock which have superior natural growth potential. Recording plays an essential role in this selection process, as shown in Figure 8.1.

Figure 8.1   **Relationship between recording and selection of breeding stock**

Both cow and bull performance are so interrelated that a full recording programme is worthwhile in order to present a detailed summary of the breeding stock involved. Records should include information under the following headings:

- Dam Ear Number
- Sire Code Number
- Calving Date
- Calving Code (denoting ease of calving)
- Birth Weight
- Calf Ear Number
- Calf Sex
- Weaning Weight
- Weaning Date
- Sale Date
- Sale Weight
- Sale Price

It can be seen from these headings that a full picture can easily be built up of the breeding values of individual cows and bulls. Where two or more herd sires are used, comparative progeny evaluation of each bull can be carried out, bringing factual accuracy into sire selection. This system of recording also identifies those cows which need culling for consistently producing inferior calves.

Chapter 9
The Suckler Herds at Givendale Farm

Background

Givendale Farm was acquired in 1974 by the JSR Farming Group. The farm had been badly neglected so it was necessary to devise a sound management plan which would fit the resources of the farm and which would also achieve a worthwhile return on tenants' capital. The farm plan was dictated by the fact that one third of the area was steep, unploughable grassland. It was also recognised that the poor quality arable land needed to be improved if cereal yields were to reach profitable levels.

The financial viability of Givendale clearly hinged on the development of a system which would result in the successful integration of both livestock and arable enterprises. So the decision was taken to utilise the permanent grassland with suckler cows and a breeding ewe flock. Winter forage for both enterprises would be produced from straw and 2-year leys, grown as a break crop in the arable rotation.

The farm, which extends to 402 hectares, is situated on the western edge of the arable Yorkshire Wolds. The arable land is cropped with cereals and grows 2-year leys. It comprises stoney, chalk wold, which is mainly sloping and exposed, rising 213 metres above sea level. The unit is now split into 271 hectares of productive arable land and 131 hectares of unploughable permanent grassland and steep dale. A plan of the farm is given in Figure 9.1 with a profile showing the topography in relation to land use in Figure 9.2.

After two years of considerable pasture improvement the permanent grass, in conjunction with 60 hectares of two-year leys carried 150 Aberdeen Angus cross Friesian suckler cows, (80 calving in the autumn and 70 calving in the winter and 850 mule ewes). In 1980 a change of policy led to a reduction in the ewe flock to 650 ewes. This released 12 hectares of temporary grass for more profitable arable production. At the same time the pedigree Charolais herd was being established to produce reliable, high-performance bulls for use on the suckler herd and for sale. This policy was designed to add value to the cattle enterprise.

Figure 9.1   Plan of Givendale Farm showing the grazing, arable and wooded areas

Key

⌘ Wooded areas

▧ Permanent pasture divided into paddocks (131 hectares)

☐ Arable land (271 hectares) and roads

For profile from A to B (1.5 miles) see Figure 9.2.

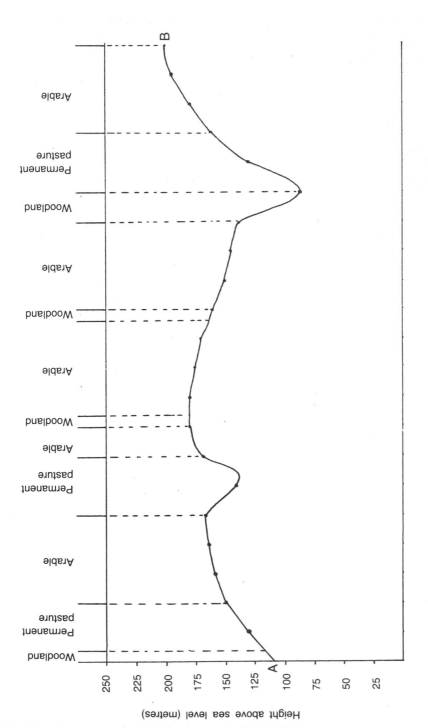

Figure 9.2 Topography of Givendale Farm in relation to land use. Profile from A to B (see Figure 9.1)

A herd of Angus x Holstein Friesians grazing permanent pasture at Givendale.

A group of suckled heifer calves. The result of planned breeding and a tight calving pattern.

The increasing costs associated with feeding autumn-calving suckler cows led to the sale of the herd in 1993. At that time the winter-calving herd was increased from 70 to 100 cows and the Charolais herd started to expand to its present level of 110 breeding cows.

In 1990 I formed the GLB Charolais Improvement Group by inviting two other breeders to co-operate in a breeding programme, the aim of which is to use BLUP Technology to achieve fast-track genetic improvement in a combined population of 250 Charolais cows.

The development of the suckler herds at Givendale has been built on strict management controls which were intended to maximise performance at every stage of production. Particular emphasis has been put on maintaining a tight 6-week calving period for each herd and on achieving good calf performance. The progress of the herds has been kept on target with the help of the MLC Beefplan recording service.

The Grazing System

When the farm was acquired the grassland was heavily infested with rabbits, and the presence of large areas of scrub considerably reduced the area of productive grazing. In addition, much of the fencing was dilapidated and had to be replaced. The improvement plan for the farm emphasised the need to maximise the stock-carrying capacity of the grassland, with ease of management a priority.

The scrub and the rabbit burrows were bulldozed and the areas of grassland were re-fenced to create paddocks of about 12 hectares in size, as shown in Figure 9.1. Some reseeding of the areas which were damaged by the bulldozers was necessary but, in the main, pasture improvement has been achieved by encouraging the indigenous grasses with the use of fertilisers, sprays and close grazing. The division of the grassland into 12-hectare paddocks has enabled a controlled, clean grazing system to operate by alternating grazing between the cattle and sheep every other year. The paddock system has also proved essential in the establishment of bull beef production from the suckler herds.

Groups of cattle and sheep are set-stocked at about 1 cow and calf per acre or 5 ewes and lambs per acre. The intensity of stocking varies slightly, depending on the grass production potential of each area. Fertilisers have played an important role in improving pasture productivity. The first dressing of nitrogen fertiliser (90 kg/ha) is applied in early March. The second application of 90 kg per hectare is spread at the end of May. A further 40 kg per hectare is applied at the end of June, and a 15-15-15 compound is then applied in early August at the rate of 90 kg per hectare to maintain the P and K levels and to ensure autumn grass growth.

Weed control in the grassland is considered very important. Chemical sprays are used with care to control problem weeds such as nettles and thistles.

Forage

Temporary 2-year leys provide silage for the cows as well as grazing for the weaned lambs from the ewe flock. Leys are established by drilling into a properly ploughed and worked seed bed, after winter barley in early August. Tetraploid ryegrasses are grown and three cuts of silage are taken before the end of July, when lambs begin to graze the leys. A total of 375 kg of nitrogen is applied per hectare during the season. The grass is cut with a trailed mower-conditioner and is then picked up after a 24-hour wilt with a precision chop machine.

The target is to make material with a 'D' value of 67, dry matter (DM) of 25 per cent and an energy content of 10.8 MJ ME/kg DM.

The other form of forage fed at Givendale is wheat straw treated with ammonia. This has proved to be an excellent feed for dry cows replacement heifers and pedigree Charolais cows. The availability of large quantities of wheat straw, coupled with economic treatment of 30-tonne stacks has made the product a useful alternative to silage.

Season of Calving

Due to the availability of 2-year leys it was decided that a high input, high output, silage-based suckler system would best fit the resources of the farm. Further considerations, such as the types of buildings that were in existence on the farm, contributed to the decision to calve 80 cows in the autumn and 70 cows in the winter. In addition, by splitting the calving season the same herd sires could be used twice in one year. This policy changed in 1993, when emphasis moved to winter calving.

Calves sired by Givendale Gascon, a national trait leader for Beef Value.

Cow Type

Aberdeen Angus cross Friesian cows were chosen because of their breeding efficiency, their milkiness and because of their ability to produce fast-growing, good-quality calves. However, in recent times, the Holstein influence has made inroads into the breed and is having detrimental effects on herd performance. To counter these effects a new replacement strategy is in place to develop a composite breed of cow, as described in Chapter 1.

Culling and Herd Replacements

Cows are culled when they are 10 years old or when they develop physical problems such as infertility, or diseased or pendulous udders. The herd is replaced at a rate of 14 per cent per year. Heifers, selected and bought as yearlings, are mated when they weigh at least 350 kg, using a Limousin bull. They calve down 4 weeks before the cows, allowing more time for postcalving recovery and giving more time before weaning for their smaller calves to grow. They are always housed separately to allow better control of feeding in order to bring them to the correct condition for successful re-breeding.

Herd Sires

The Limousin bull used on the heifers was selected with a positive Calving Value (CV) to ensure the easy calving of his progeny.

All the cows are mated to Charolais bulls because, as discussed in an earlier chapter, calves sired by a Charolais grow faster and reach heavier weights than calves sired by other breeds. Bulls are selected from the Givendale pedigree Charolais herd using individual trait EBVs and the Beef Value to identify the top sires. Emphasis is placed on good growth and muscle scores, while, at the same time, keeping an eye on the Calving Value.

The Autumn Calving Herd

For the benefit of those farmers who still run autumn-calving herds the following account describes the management of the Givendale herd to the point when it was sold in 1993.

Cows calved outside from mid-September to the end of October, with 85 per cent of the herd calving in the first 4 weeks. Shortly before calving was due to start, all the cows were moved into an 8 hectare paddock and magnesium syrup was made available to prevent grass staggers. The herd was observed every 3 hours around the clock, with night patrols carried out using a Land Rover equipped with a spot light. A careful watch was kept on calving cows, and assistance was given when necessary in a wooden catching pen constructed in the calving field. A calving aid was always available.

The calves were tagged soon after they were born to correspond with their mothers' numbers. They were then moved into adjacent paddocks where they remained until they were housed. The herd stayed out until the end of October and straw was usually fed to the cows in order to maintain dry matter intake. When the cows and calves were housed, the yarding system allowed the grouping together of replacement heifers, the second calvers and thin cows, and the fitter mature cows. Yarding the herd in these 3 groups enables feeding to be regulated according to the nutritional requirements of the differing classes of cow.

The herd was fed silage *ad libitum* with increasing quantities of rolled barley up to the 6 week mating period, by which time the heifers and the younger cows were receiving 2.5 kg barley per day and the mature cows 1.8 kg per day. In addition, a high phosphorus mineral (10% P) was rationed by sprinkling it onto the silage. This feeding regime was intended to provide the cows with a rising plane of nutrition to ensure that good conception rates were achieved. The target was for about 96 per cent of the cows to hold in calf during the 6 week mating period. To achieve this standard the cows must not only be well-fed but they must also be healthy, so they were wormed towards the end of November with a broad spectrum anthelmintic; they were also treated for lice.

In February the cows with the bull calves were split from the cows with heifer calves and they stayed as separate groups during the grazing and finishing periods. During February the high phosphorus mineral was replaced by a general purpose one and the daily ration of barley was slowly reduced. The rate of reduction depended on the condition of the cows and on the quality of the silage.

The calves were creep-fed during the winter, up to a maximum of 2.5 kg per head per day. The creep feed was a mixture of rolled barley and sugar beet pulp, with a protein balancer added. The target weights for the calves by the time they were turned out to grass were 320 kg for the bulls and 290 kg for the heifers.

Before the herd was turned out during the last week in April, the cows were pregnancy tested. Any barren cows were either re-mated to come into the winter calving herd or they were earmarked for culling after they had put on condition at grass. Barren cows were never turned out with the cows which had bull calves at foot. When the herd was turned out to grass the cows and calves were set stocked on the grazing areas. Liquid magnesium syrup was made available to the cows in ball feeders for 6 weeks, to counteract the risk of hypomagnesaemia.

The calves grew well at grass, with the heifers gaining about 1 kg per day and the bulls about 1.2 kg a day. These gains were achieved on grass and milk only. When the calves were weaned at the end of July they were weighed, wormed and then housed for finishing.

The cows were treated with a dry-cow antibiotic, their teats were dipped and they were then housed for 3 days and fed on straw. Liquid magnesium syrup was also fed at this stage to help the cows to digest the straw and to counteract the possibility of hypomagnesaemia associated with stress. When the cows were turned out again they were moved onto bare pasture, at a high stocking rate. This prevented them from becoming too fat before they started to calve again in the middle of September.

The management of the autumn calving herd over the year is summarised in Figure 9.3.

The Winter Calving Herd

The dry cows are brought inside when the weather deteriorates in November. They are loose-housed in straw-bedded yards and fed anhydrous ammonia-treated wheat straw supplemented with a special high sulphur mineral. The second calvers are usually penned separately and fed 1.5 kg of rolled barley per day to improve their body condition. The cows are wormed with a broad spectrum anthelmintic and they are also treated for lice. All the cows and heifers are vaccinated to prevent their calves from contracting rotavirus or E. coli infections. Vaccination takes place about a month before calving is due and the preventative programme has proved to be very successful.

As the cows start to calve, their nutritional requirements increase so silage replaces the treated wheat straw. Calving is monitored very carefully and the herd is seen every three hours round the clock. The cows are penned in groups of twenty-five as they calve, and the yards are kept well bedded with clean straw.

As soon as the calves are born their navels are soaked in strong iodine solution and the treatment is repeated the next day. Iodine has proved to be a good material for sterilising and drying the umbilical cord quickly, preventing navel infections. It is, of course, essential for every calf to get the right amount of colostrum within six hours of birth and all calves are checked to make sure that they suckle their mothers properly. The calves are watched very carefully for any signs of scouring or pneumonia so that treatment can be implemented swiftly if necessary. We aim to rear healthy calves that will grow to their full potential unhindered by disease.

Figure 9.3   The autumn calving herd calendar

MAY	Cows and calves set stocked in grazing areas			Magnesium syrup available
JUNE				
JULY				
AUGUST	Calves—weaned, wormed, weighed. Cows—dry cow therapy			
SEPT	Heifers calve	Cows stocked tightly to prevent them getting over fat	Magnesium syrup fed to heifers	Magnesium syrup fed to cows
OCT	Cows calve under close supervision. Calves tagged and weighed	85% of the herd calves in 4 weeks	Feeding based on silage. Cows fed an increasing amount of rolled barley up to mating period when they get 2.5 kg	Calves are offered creep feed at housing and by mid January they are eating 2 kg /head/day. This level of feed is maintained until turnout
NOV		Cows and calves housed		
		Calves de-horned		
		Cows wormed and treated for lice		
DEC	Mating period—careful watch kept on bull performance. Serving dates recorded	Cows receive high phosphorus mineral		
JAN				
FEB	Herd split—cows with bull calves separated from cows with heifer calves		Barley ration slowly reduced to 1 kg/head /day	
MARCH				
APRIL				
	Turn out—calves weighed, cows pregnancy diagnosed			

The cows are fed on good quality silage, which is sufficient to provide for maintenance and lactation, until they are turned out to grass at the end of April. Heifers, however, are fed 2 kg per day of rolled barley in addition to silage. Before turn-out the calves are dis-budded and both the cows and the calves are given copper needles to supplement copper-deficient pastures. A high phosphorus mineral is given to the cows three weeks prior to the start of the mating period, which coincides with the herd being turned out to grass. The herd is split into two groups and a bull is run with each herd for three weeks, after which time the herds are joined together and a third bull is turned in for a further three weeks. To counteract the risk of hypomagnesaemia, magnesium syrup is made available to the cows for the first six weeks of the grazing season.

In early July the cows with the bull calves are separated from the cows with the heifer calves, and the two groups are then grazed in different paddocks. At this stage creep feed is made available to the calves, in hopper feeders, allowing one five-space feeder for 20 calves. Initially the calves take small quantities of creep but by the end of August it needs to be restricted to 2.5 kg per calf per day. If grass runs short in the autumn the cows are buffer-fed with either spring barley straw or ammonia-treated straw. Magnesium syrup is fed to the cows from about 1 September until the calves are weaned in the middle of October.

At weaning, the calves are weighed and wormed, then housed for finishing. Dry cow therapy is practised and the cows are pregnancy tested. Those found not to be in calf are culled to comply with retention periods. After a housed, three-day drying off period the in-calf cows are turned out again for about another month before they are housed for the winter.

Figure 9.4 summarises the annual calendar for the winter calving herd.

Improvements in Performance

Table 9.1 shows the average calf weaning weights which were achieved at Givendale during the 10-year period from 1984 to 1993. The improved performance of the calves was the result of putting into practice the management techniques which have been outlined in this book.

Figure 9.4   **The winter calving herd calendar**

MAY	Cows and	Mating period –	Magnesium
	calves	careful watch on	syrup fed
JUNE	set	bull performance	
JULY	stocked	Cows with bull	
	in	calves are split	
AUG	grazing	from cows with	
	areas	heifer calves.	
SEPT		Creep feed	Magnesium
OCT		fed to calves	syrup fed

Calves—weaned, wormed, weighed and housed
Cows—dry cow therapy and pregnancy diagnosed

NOV	Cows outside		
DEC	Cows housed	Cows and heifers wormed	
		Heifers vaccinated with Rotavac	
	and fed	Cows vaccinated with Rotavac	
JAN	ammonia treated		
	wheat straw	Heifers calve	Health status of calves watched carefully.
FEB			Particular attention paid to
	Cows fed silage and a high-phosphorus mineral. Copper needles are given before turn-out	Cows calve under close supervision. Calves tagged and weighed	— colostrum — navel dressing — scours — pneumonia. Copper needles given and disbudding carried out before turn-out
MARCH			
APRIL			

The Future

As the downward pressure on margins continues, suckled calf producers must become more sensitive to the benefits of breeding technology to reduce costs of production and to improve efficiency. To help achieve these goals we now have the means to identify superior breeding stock using EBVs.

One of the major challenges facing the industry is the organisation of effective replacement policies. The few remaining research centres have unfortunately not kept pace with events and have relied solely on dairy-bred cows for their research projects. This has left producers starved of guidance as they embark on unproven crossing programmes which have no long-term direction. Breeding projects by nature are long-term so it is essential that research in this field is urgently pursued to develop replacement strategies which deliver a high-quality, uniform animal efficiently and competitively.

Suckled calf production will remain in a financially precarious state as the finished beef market remains close to world prices. Those continuing in the industry will tend to be larger producers who control overhead costs and fine-tune their technical ability to reduce the unit costs of production to maximise returns. Even these efficient producers will rely heavily on the relevant EU premium schemes and extensification payments.

Table 9.1   Average calf weaning weights (kg) at Givendale from 1985 to 1994

	1985	1986	1987	1988	1989	1990	1991	1992	1993	1994
Autumn born weaned at 10 months										
Charolais X bulls	390	418	20	425	430	442	448	451	455	-
Charolais X heifers	349	354	351	354	352	360	365	368	371	-
Winter born weaned at 7 months										
Charolais X bulls	309	312	320	335	338	341	348	352	360	362
Charolais X heifers	270	265	275	268	271	283	297	301	312	317

Table 9.2   **Gross margins (£) recorded by the MLC at Givendale from 1990 to 1994 for the winter calving herd**

	1990	1991	1992	1993	1994
Gross margin / cow	306	343	364	375	380
Gross margin / hectare	624	631	640	651	667

The increase in calf performance is reflected in the improvement in both the gross margin figures.

In this book I have outlined the management strategies, which have been developed at Givendale, to maximise herd output. The future cow herd will be developed using an imported four-breed composite maternal dam line from the USA which was developed at the Clay Research Centre in Nebraska. This cow-type has been proven to be highly efficient, when run on large-scale commercial units.

The Givendale Charolais herd of 110 cows will continue to work within GLB Charolais Ltd. to produce high-performance sire-line bulls. The main focus of selection concentrates on muscle score and growth traits to increase efficiency.

Castle Learning Resource Centre